Remote Sensing of Large Wildfires
in the European Mediterranean Basin

Springer-Verlag Berlin Heidelberg GmbH

Emilio Chuvieco (Ed.)

Remote Sensing of Large Wildfires

in the European Mediterranean Basin

With 45 Figures, 16 Color Plates and 33 Tables

 Springer

Prof. Dr. Emilio Chuvieco

University of Alcalá
Department of Geography
Colegios, 2
28801 Alcalá de Henares
Spain

E-mail: emilio.chuvieco@uah.es

Additional material to this book can be downloaded from http://extras.springer.com.

ISBN 978-3-540-66464-2   ISBN 978-3-642-60247-4 (eBook)
DOI 10.1007/978-3-642-60247-4

Library of Congress Cataloging–in–Publication Data
Remote sensing of large wildfires in the European Mediterranean Basin / Emilio Chuvieco (ed).
Includes bibliographical references.
1. Forest fires–Mediterranean Region–Prevention and control.
2. Wildfires–Mediterranean Region–Prevention and control.
3. Remote sensing.
4. Geographic information system.
I. Chuvieco Salinero, Emilio.
SD421.32.M44R46   1999
634.9 ' 618 ' 028--dc21   99-15543 CIP

© Springer-Verlag Berlin Heidelberg 1999
Originally published by Springer-Verlag Berlin Heidelberg New York in 1999

Production: ProduServ GmbH Verlagsservice, Berlin
Typesetting: Camera-ready by editor
Cover design: Erich Kirchner, Springer Heidelberg
SPIN: 10686078   32/3020-5 4 3 2 1 0 - Printed on acid-free paper

Foreword

Forest fires are considered a major environmental problem in many European Union Member States as well as in other parts of the world. According to a recent report of the European Commission, forest fires are a dominant feature of the landscapes of the five Southern European Member States - Portugal, Spain, France, Italy and Greece, and almost a half of the Community's forests have been classed as fire-risk areas.

Protection from forest fires is an interdisciplinary endeavour, which needs to be addressed from both the technological and methodological point of view, and which necessitates a wide spectrum of various scientific disciplines. It also implies the solution of numerous practical problems (both of a generic nature and of a specific one) and the consideration of the characteristics of the specific areas in which fire fighting is carried out.

Scientific research is offering a major contribution to forest fire fighting. The European Commission (DG XII, Directorate General for Science, Research and Development) has supported since the 1980s a number of multinational research projects through its successive RTD programmes in the field of the Environment. Findings from these integrated and multidisciplinary projects have increased our understanding of the causes and socio-economic aspects of fire, of the natural role of fires in the maintenance and restoration of ecosystems, and of fire behaviour modelling. Emphasis was also given to the development and validation of fire behaviour and fire fighting models, fire danger indexes and to improving the scientific basis of the new technologies to develop tools in support of fire management. Special effort has been made to involve potential users of research results to improve the dissemination and exploitation of the results.

Earth observation from space has indeed a significant role to play in the field of forest fires, covering some of its important scale aspects, in terms of time and space. Observations from space need to be validated and checked by field studies, and data generated from space must be used in models to provide useful information for operational applications. The rapidly advancing technology of space observation gives a unique opportunity for the solution of the forest fire problem. The use of satellites of the new generation, together with advanced information technologies, and the collaboration of fire modelling experts, fire managers and remote sensing specialists, can provide alternative solutions to current operational aspects of the fire problem, and are certain to be adopted rapidly by fire management entities.

This volume encompasses studies carried out under the MEGAFiReS (Remote Sensing of Large Wildland Fires in the European Mediterranean Basin) project, funded by DG XII, to help gain insight into the potential role of the use of remote sensing techniques in the field of forest fire detection, monitoring and assessment. Focusing on areas like those in the Mediterranean, where early fire detection and reliable fire danger prediction can not only help cut costs, but help to reduce the yearly burned area and to avoid casualties, this volume provides a comprehensive report of the current status of European research into the strengths and limitations of currently available satellites for operational fire management.

Anver Ghazi and Panagiotis Balabanis
Global Change and Biodiversity Unit
European Commission, Brussels.

Contents

List of contributors

Inmaculada Aguado
Research assistant, Department of Geography, University of Alcalá
Colegios, 2; 28801 Alcalá de Henares, Spain.
ggias@geogra.alcala.es

Giovanni Bovio
Department AGROSELVITER, University of Turin
Via L. Da Vinci 44; I-10095 Torino, Italy.
giovanni.bovio@agraria.unito.it

Andrea Camia
Assistant Professor, Department AGROSELVITER, University of Turin
Via L. Da Vinci 44; I-10095 Torino, Italy.
andrea.camia@agraria.unito.it

João M.B. Carreiras
Research Assistant, Departamento de Engenharia Florestal,
Instituto Superior de Agronomia
Tapada da Ajuda, 1300 Lisboa, Portugal.

Luis Carvacho
Associate professor, Institute of Geography, Catholic University of Chile
Avda. Vicuña Mackenna, 4860, Santiago, Chile.
lcarvach@drumlin.geo.puc.cl

Pietro Ceccato
Researcher, Global Vegetation Monitoring Unit, Space Applications Institute,
Joint Research Centre, European Commission, 21020 Ispra (Va), Italy.
pietro.ceccato@jrc.it

Emilio Chuvieco
Professor, Department of Geography, University of Alcalá,
Colegios, 2, 28801, Alcalá de Henares, Spain.
Emilio.chuvieco@uah.es

David Cocero
Research assistant, Department of Geography, University of Alcalá
Colegios, 2; 28801 Alcalá de Henares, Spain.
ggdcm@geogra.alcala.es

Michel Deshayes
Researcher. CEMAGREF - ENGREF Remote Sensing Laboratory
500 rue J.F. Breton, 34 093 Montpellier Cedex 5, France
deshayes@teledetection.fr

Ian Downey
Principal Scientist, Natural Resources Institute,
The University of Greenwich, Medway University Campus,
Chatham Maritime, Kent ME4 4TB, UK
i.downey@greenwich.ac.uk

Arturo Fernández-Palacios
Head of the Remote Sensing Unit, Environmental Agency of Andalucía
Avda. Leonardo da Vinci s/n, Cartuja 93, 41071 Sevilla, Spain.
Dp.Teledeteccion@cma.junta-andalucia.es

Stéphane Flasse
Principal Scientist, Natural Resources Institute,
The University of Greenwich, Medway University Campus,
Chatham Maritime, Kent ME4 4TB, UK
s.p.f.flasse@greenwich.ac.uk

Jesús Jurado
Technician, Environmental Agency of Andalucía
Avda. Leonardo da Vinci s/n, Cartuja 93, 41071 Sevilla, Spain.
Dp.Teledeteccion@cma.junta-andalucia.es

Michael Karteris
Department of Forestry and Natural Environment, Aristotelian University,
Box 248; GR-54006 Thessaloniki, Greece.
karteris@for.auth.gr

Nikos Koutsias
Department of Forestry and Natural Environment, Aristotelian University,
Box 248; GR-54006 Thessaloniki, Greece.
koutsias@for.auth.gr

A. Lobo
Instituto de Ciencias de la Tierra Jaume Almera
Consejo Superior de Investigaciones Científicas
C/ Lluis Solé Sabaris s/n, 08028 Barcelona, Spain.
alobo@ija.csic.es

M. Pilar Martín
Assistant Professor, Department of Geography, University of Alcalá
Colegios, 2, 28801, Alcalá de Henares, Spain.
ggmpmi@geogra.alcala.es

Carmen Navarro
Technician, Environmental Agency of Andalucía
Avda. Leonardo da Vinci s/n, Cartuja 93, 41071 Sevilla, Spain.
Dp.Teledeteccion@cma.junta-andalucia.es

Rafael Navarro
Associate Professor, Department of Rural Engineering, University of Córdoba
Avda. Menéndez Pidal s/n, 14080 Córdoba, Spain.
ir1nacer@lucano.uco.es

Juli G. Pausas
Senior Researcher, Centro de Estudios Ambientales del Mediterráneo (CEAM)
Parc Tecnològic, 46980 Paterna, Valencia, Spain.
juli@ceam.es

José M.C. Pereira
Associate Professor, Departamento de Engenharia Florestal,
Instituto Superior de Agronomia
Tapada da Ajuda, 1300 Lisboa, Portugal.
jmcpereira@isa.utl.pt

David Riaño
Research assistant, Department of Geography, University of Alcalá
Colegios, 2; 28801 Alcalá de Henares, Spain.
gargamel@cestel.es

Francisco Rodríguez-Silva
Head of the Forest Fire Unit, Environmental Agency of Andalucia,
Avenida de Eritaña, 2; ES-41071 Sevilla, Spain.
frysilva@arrakis.es

Ana C.L. Sá
Research Assistant, Departamento de Engenharia Florestal,
Instituto Superior de Agronomia
Tapada da Ajuda, 1300 Lisboa, Portugal.
anasa@drag.isa.utl.pt

F. Javier Salas
Assistant professor, Department of Geography, University of Alcalá
Colegios, 2; E-28801 Alcalá de Henares, Spain.
ggjsr@geogra.alcala.es

Teresa N. Santos
Research Assistant, Departamento de Engenharia Florestal,
Instituto Superior de Agronomia
Tapada da Ajuda, 1300 Lisboa, Portugal.

João M.N. Silva
Research Assistant, Departamento de Engenharia Florestal,
Instituto Superior de Agronomia
Tapada da Ajuda, 1300 Lisboa, Portugal.

Adélia M.O. Sousa
Research Assistant, Departamento de Engenharia Florestal,
Instituto Superior de Agronomia
Tapada da Ajuda, 1300 Lisboa, Portugal.
adeliasousa@isa.utl.pt

Nicolas Stach
Research Assistant, CEMAGREF - ENGREF Remote Sensing Laboratory
500 rue J.F.Breton; 34093 Montpellier, Cedex 5, France.
mouss@teledetection.fr

V. Ramon Vallejo
Head of the Restoration Program, Centro de Estudios Ambientales del
Mediterráneo (CEAM)
Parc Tecnològic, 46980 Paterna, Valencia, Spain.
ramonv@ceam.es

1   Introduction

Emilio Chuvieco
Department of Geography, University of Alcalá (Spain)

Wildland fires are becoming a major concern for several Environmental Sciences. Assessment on fire effects on a local scale is increasingly considered a critical aspect of ecosystem functioning, since fire plays a crucial role in vegetation composition, biodiversity, soil erosion and the hydrological cycle. On a global scale, fire is the most generalised means to transform tropical forest in agricultural areas, and it has severe impacts on global atmospheric chemistry.

Fire is a natural factor in many climates, such as the Mediterranean, with high levels of vegetation stress during the summer season. However, changes in traditional land use patterns have recently modified the incidence of fire in this area. Rural abandonment in the European Mediterranean Basin has implied an unusual accumulation of forest fuels, which notably increases fire risk and fire severity. On the other hand, the increasing use of forest as a recreational resource involves a higher incidence of man-induced fires, either by carelessness or arson.

In spite of the great incidence of fires in Southern Europe, a significant amount of information is still required to better understand fire risk factors and fire effects. Most of the national forest services do not provide cartographic representation of burned areas. Therefore, there is a lack of understanding about the spatial factors related to fire incidence and the spatial consequences of fires. For instance, on many occasions only a general estimation of burned area is provided, but the fire perimeter is not available, and thus the fire fighting manager does not clearly know the spatial pattern of fire behaviour or the areas more severely affected by the fire. In the case of fire risk estimation, most of the current danger indices are based on weather stations, which are frequently sparse and located far from the forested areas. Consequently, only general information about the spatial distribution of risk is available, and fire pre-suppression resources might not be optimally allocated.

Remote sensing from space is especially suitable for forest fire-related research. The wide area coverage and high frequency provided by satellite sensors, as well as their information on non visible spectral regions, makes them a very valuable tool for prevention, detection and mapping of wildland fires. During the last decade, the range of applications has significantly increased, making satellite remote sensing a solid ally in many forest fire strategic plans.

This book presents and in-depth review of remote sensing applications to forest fire research, with several operative examples centred on the European Mediterranean Basin, most of them taken from the European project Megafires. The contents are organised in three major sections, dealing with the three phases of fire management: pre-fire planning (fire risk, Chaps. 3 to 5), fire suppression (detection and fighting, Chap. 6) and fire effects (burnt land mapping, Chaps. 7 to 9). An introductory section on the ecological impacts of fires in the Mediterranean environment precedes these chapters, in order to provide a framework to better understand the natural scenario of this research.

To help the reader, the different chapters are illustrated with a wide range of graphs and satellite images. All the colour figures are included in a separate section, located in the centre of the book. For educational purposes and to stimulate further research in this field, a CD-ROM has been additionally produced for the book. It includes samples of satellite images, programs and digital maps generated within the Megafires project. All these files are easily accessible from external software, although some display utilities have also been included. All the components of the Megafires consortium are convinced of the necessity of disseminating, as much as possible, work financed from public sources (in this case, the Environment and Climate Program of the European Union DG-XI). The object of this is to encourage other teams to take advantage of our results to continue with their research.

We hope this book can contribute significantly to reduce the negative effects of fire on the social and natural environment, in the Mediterranean and elsewhere.

2 The role of fire in European Mediterranean ecosystems

Juli G. Pausas and V. Ramón Vallejo
Centro de Estudios Ambientales del Mediterráneo (CEAM), Valencia (Spain).

Abstract. Fire is an integral part of many ecosystems, including the Mediterranean ones. However, in recent decades the general trend in number of fires and surface burnt in European Mediterranean areas has increased spectacularly. This increase is due to: (a) land-use changes (rural depopulation is increasing land abandonment and consequently, fuel accumulation); and (b) climatic warming (which is reducing fuel humidity and increasing fire risk and fire spread). The main effects of fire on soils are: loss of nutrients during burning and increased risk of erosion after burning. The latter is in fact related to the regeneration traits of the previous vegetation and to the environmental conditions. The principal regeneration traits of plants are: capacity to resprout after fire and fire stimulation of the establishment of new individuals. These two traits give a possible combination of four functional types from the point of view of regeneration after fire, and different relative proportions of these plant types may determine the post-fire regeneration and erosion risk. Field observations in Spain show better regeneration in limestone bedrock type than in marls, and in north-facing slopes than in south-facing ones. Models of vegetation dynamics can be built from the knowledge of plant traits and may help us in predicting post-fire vegetation and long-term vegetation changes under recurrent fires.

2.1 Introduction

There are several features that make the landscapes of the European Mediterranean Basin different from those of the rest of Europe, and these differences are mainly related to the climate, the long and intense human impact, and the role of fire. The latter is, in turn, influenced by the other two. Mediterranean ecosystems of Europe have been subjected to a long-term history of human use (Wainwright 1994, Grove 1996, Margaris et al. 1996), and this has provoked an older and very intense disturbance regime when compared to the other Mediterranean-climate regions in the world (Fox and Fox 1986). However, within the Mediterranean basin, differences in land-use patterns have increased during this century between Euro-Mediterranean and Afro-Asiatic-Mediterranean countries (Blondel and Aronson 1995). In the southern areas of the western Mediterranean (Maghreb), growing populations are reducing forests and shrublands by overgrazing and extending arable lands, whereas in the northern countries abandoned land is increasing at the expense of marginal agriculture (Puigdefábregas and Mendizabal 1998). These differential trends make the European Mediterranean Basin more fire prone than the southern area, as shown in the fire statistics of the last decades (Vélez 1997, Moreno et al. 1998).

2.2
Fire history

2.2.1
Statistics

Natural fires are common in many parts of the world and are an integral part of many terrestrial ecosystems. Fire has been used by man as a management tool since early times. It has been suggested that Palaeolithic people already burnt deliberately to facilitate hunting and food gathering (Stewart 1956). The first evidence of human-induced changes by fire in the Mediterranean landscape is during the Neolithic (Naveh 1975). Since then, the Mediterranean basin has seen the evolution of many cultures, some with high population densities, and most making use of fire and farming. However, from the 60s until today the general trend in number of fires and surface burnt in the European Mediterranean areas (mainly Iberian, Italic and Greek Peninsulas and surrounding islands) has increased exponentially. Fire statistics compiled for Spain from the 60s (Martínez-Ruiz 1994, Moreno et al. 1998, Piñol et al. 1998) show a clear increase in number of fires and surface burnt especially since the mid-70s (Figs. 2.1 and 2.2). From 1960 to 1973 the mean annual burnt area was about 50 kha and the annual number of fires was less than 2000. However, since 1974, the mean annual area burnt is about 215 kha, caused by a mean of 8550 annual fires, and in some of these years the area burnt was more than 400 kha, i.e. nearly 2% of the total non-arable land of Spain (1978, 1985, 1989 and 1994; Fig. 2.1). This increase has occurred in spite of the increased fire suppression efforts of the recent years. A similar trend has been found in other European Mediterranean areas (e.g. Kailidis 1992, Viegas 1998), although the increment may be slightly shifted in time. The year 1993-94 was a turning point in recent fire history because several large fire episodes took place in different Mediterranean ecosystems of the world (SE Australia: January 1994, Spain: August 1994, California: October 1993; Moreno 1998).

Data from Spain (Martínez-Ruiz 1994, Moreno et al. 1998) show that the increase in the number of fires has mainly affected non-wooded areas (e.g. shrublands); wooded areas showed lower increase (Fig. 2.1). Most areas burnt are pine forests, especially *Pinus pinaster* and *P. halepensis* (Fig. 2.3), while broadleaf species represent a small proportion. Although in Spain a large percent of the causes of fire ignition are unknown (38%), most fires are caused by people, either intentionally, or by negligence or pasture burning. Only a very few are natural (i.e. lightning, 5% of the fire with known cause; Fig. 2.3). A more detailed analysis of the causes of fire ignition in Eastern Spain has been underway since 1995 by the local government. This study has determined the origins of 99% of fires in the area. It shows that during the period 1995-97 the pattern was quite different from

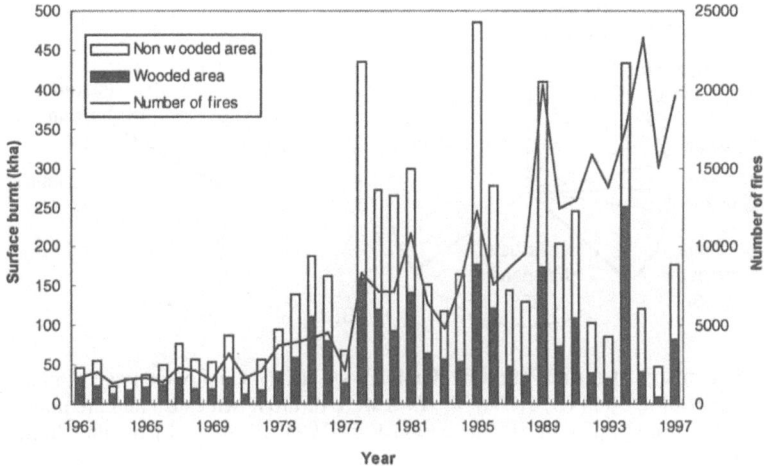

Fig. 2.1. Yearly number of fires (line), wooded surface burnt (close bars, kha) and non-wooded (e.g. shrublands) surface burnt (open bars, kha) in Spain for the period 1961 - 1997. Elaborated from data of ICONA (Martínez Ruiz 1994, Vélez 1996, 1997b).

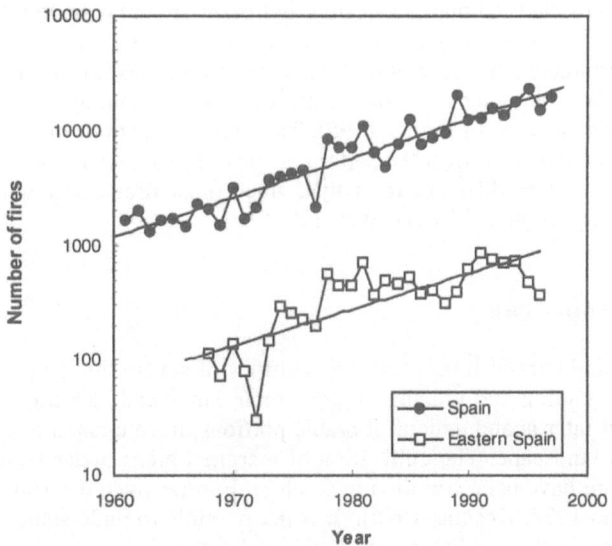

Fig. 2.2. Number of fires in the last decades for the whole Spain (black dots) and for the eastern part (Valencia region). Note the logarithmic scale of the y-axis.

the general pattern found for the whole Spain: 43% of the fires were caused by negligence, 28% were intentional and 23% were started by lightning. This elevated figure for lightning as compared with the whole of Spain may be due to the relative high frequency of dry storms inland in early summer in the region. However, it should be taken into account that this information is only available for 3

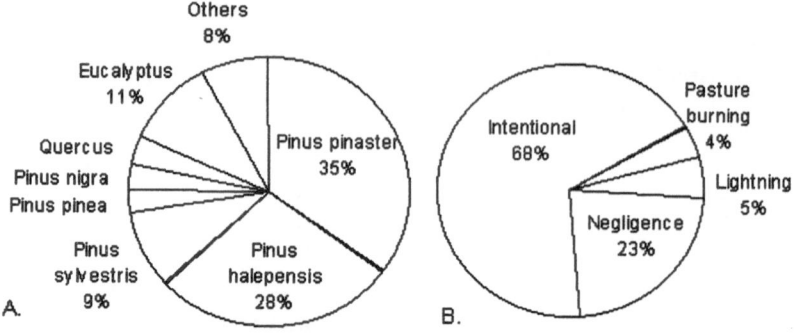

Fig. 2.3. Percentage of surface burnt by dominant species (A) and percentage of number of fires by causes of ignition (B, 38% of the fires were of unknown cause and are not considered in the figure), in Spain during the period 1974-1994. Elaborated from data in Moreno et al. (1998).

years and these years were moist years (with relatively high rainfall). This implies that the percent fires started by lightning may be high (because of the high number of storms) but the total number of fires and burnt area may be low (because of the high humidity). For the whole of Spain, fires caused by pasture burning are distributed throughout the year, with a great proportion occurring during spring and fall, whereas most other fires, including those from lightning, are concentrated in summer (Vázquez and Moreno 1995). The temporal pattern of fires from pasture burning may, in part, reflect the pattern of fires of few decades ago, but the current fire regime is very different from that, since most fires occur within just a few weeks during summer (Moreno et al. 1998).

2.2.2
Land-use changes

In the Mediterranean basin, many centuries of severe human pressure resulting in burning, cutting and grazing on non-arable lands and clearing, terracing, cultivating, and later abandonment of arable portions, have created a strongly human-influenced landscape. The cultivation of marginal areas under increasing population pressure have been common in southern Europe since the 16th century (Roxo and Mourao 1995, Kosmas 1996). It is not possible to understand current vegetation patterns in the Mediterranean basin without taking into account past anthropogenic activities and land uses. Human intervention has been so strong that it is still making a significant impact on current and future vegetation patterns.

The changes in fire occurrence during the last decades closely reflect the recent socio-economic changes underway in the European Mediterranean countries (Vélez 1993, Moreno et al. 1998). With industrial development, European Mediterranean countries have experienced: depopulation of rural areas, increases in agricultural mechanisation, decreases in grazing pressure and wood gathering, and increases in the urbanisation of rural areas (LeHouérou 1993). These changes in traditional land use and lifestyles have implied the abandonment of large areas of farmland, which has led to the recovery of vegetation (García-Ruiz et al. 1996, Roxo et al. 1996) and an increase in accumulated fuel (e.g. Rego 1992). In Southern Europe, human activity has dramatically increased fire frequency as a consequence of land abandonment and tourist pressure. Piñol and Terradas (1996) found a significant relationship between population density and fire occurrence in Mediterranean areas of the Iberian Peninsula. As a consequence of these processes, landscapes are becoming homogeneous (Moreno and Oechel 1992, Sala and Rubio 1994). In summary, land-use changes produced during the present century in southern Europe are parallel to the changes in the fire regime, from being few in number and affecting small areas, to becoming very numerous and affecting large extensions every year. This trend is not observed in the southern Mediterranean basin where traditional land uses remain the major socio-economic system.

2.2.3
Climate change

Although the main reason for fire increase in the last decades is probably changes in land use, climatic factors should be considered as a contributing factor. Fires tend to be concentrated in summer when temperatures are high, and air humidity and fuel moisture are low. Predictions on climate warming in the Mediterranean basin indicate an increase in air temperature and a reduction in summer rainfall (Houghton et al. 1996). Although there is uncertainty as to the mean and variance of the precipitation changes, all predictions suggest a future increment in water deficit. These changes would lead to an increase in water stress conditions for plants, changes in fuel conditions and increases in fire risk, with the consequent increase in ignition probability and fire propagation. Analysis of past climate data already shows some of these trends (Maheras 1988, Amanatidis et al. 1993, Piñol et al. 1998). For example, the recent analysis of meteorological data from 1910 to 1994 in the eastern Iberian Peninsula (Piñol et al. 1998) shows a clear increase in temperature and potential evapotranspiration and a reduction in summer humidity. These changes are correlated to an increase in the number of fires. The climate changes that are predicted to occur in the near future as a result of releasing greenhouse gases are likely to induce increased fire risk not only in the Mediterranean area, but also in other fire-prone regions of the world (Flannigan and van Wagner 1991, Torn and Friend 1992).

2.3
Fire effects on soils

Fire effects on soils can be separated in two phases, that is, direct losses of nutrients during burning, and post-fire changes due to low vegetation cover. These two phases include different processes, and they eventually require different measures to reduce soil degradation risk.

During burning, direct nutrient losses are produced by volatilisation, mostly C, N and S, and by ash convection in the smoke column (Raison et al. 1985, Gillon and Rapp 1989, Trabaud 1994). These losses are especially relevant in high intensity fires (Little and Ohmann 1988). In addition, the vegetation and forest floor cover disappear partially or totally, and the topsoil suffers heating (Christensen 1994). These processes associated with burning operate in a time scale of seconds to weeks, the latter in the case of smouldering in small spots. The magnitude of these impacts is included in the often ambiguous term 'fire severity', which is somehow related to fire intensity, and may be defined as the residence time of a threshold temperature, e.g. 150 °C (Pérez-Ramos 1997). As fire temperature is related to fuel load and spatial structure, actions to reduce fire impact during combustion are addressed to control fuel accumulation.

The burned land has lost most of the plant and forest floor cover, therefore the ash layer and bare soil are exposed to water and wind erosion, and to soluble nutrients leaching from the ash layer during the post-fire period of low vegetation cover (DeBano et al. 1979, Khanna and Raison 1986). The time scale of this second phase is from fire extinction up to a few months to years depending on the vegetation recovery rate. Nutrient losses produced after fire could be higher than those produced during burning, especially when soil erosion is relevant. The greatest damage by fire is caused in those areas with a long dry season where organic horizons can be burnt, exposing, and perhaps affecting, the mineral soil (Trimble 1988). In Mediterranean regions, the frequent autumn rainstorms constitute a high erosion risk after summer wildfires. Thornes (1990) considered that a minimum of 30% projective plant cover is sufficient for protecting the soil against water erosion. The higher the risk of post-fire soil erosion, the higher the time required for vegetation to reach this threshold plant-cover value. In eastern Spain, critical time values varied from a few months to more than 1.5 years (Vallejo et al. 1999), and these are quite dependent on the regenerative strategy of the vegetation. Post-fire soil degradation in vulnerable soils (erodible soils, steep slopes and low plant-regeneration capacity) could be mitigated by applying emergency seeding techniques (Vallejo and Alloza 1998) that aim at enhancing a protective herbaceous cover shortly after the fire. In spite of nutrient losses produced during combustion and post-fire leaching and erosion, soil fertility increases temporarily by ash incorporation in the soil and soil biological activity enhancement shortly after the fire (Walker et al. 1986, Kutiel and Naveh 1987), although a late short-term depression in nutrient availability may appear (Ferran 1996). Soil biological activity after fires requires more than 20 years for complete recovery (Prodon et al. 1987). In addition, fire may modify mycorrhizal inocula in the soil (Torres and Honrubia 1997, Tartaglini 1992) that can limit plant regeneration. Soil properties usually increase their spatial heterogeneity because of processes associated to the fire, such as patchiness in fire severity (Pérez-Ramos 1997), short-distance redistribution of ashes and top soil material after fire (Serrasolsas 1994), and the modifications in nutrient cycling produced by the different litter qualities of colonising plants (Ferran and Vallejo 1992).

High fire frequency affecting nutrient-poor ecosystems strongly increases the risk of soil fertility depletion and desertification. In semiarid shrublands of southeastern Spain, Carreira et al. (1996) found a sharp depletion of soil-available nutrients, especially extractable inorganic phosphorus, associated to a linear increase in the fire frequency along a successional chronosequence. In addition, C accumulation in the soil could be limited by N losses during fire in ecosystems affected by high fire frequency (Menaut et al. 1993, Vitousek and Howarth 1991). In spite of the short-term nutrient losses produced by severe wildfires, and that model predictions indicate that regular burning may result in a decline of forest productivity (O'Connell 1989), little evidence has been found so far of long-term decreases in soil productivity because of fire. Recently, Ferran et al. (1998) found that 3.5 years after the fire, *Quercus coccifera* garrigues accumulated less biomass in stands affected by recurrent fires (up to 3 fires in 16 years) as compared to those affected by only one.

2.4
Post-fire regeneration of vegetation

The effects of fire on vegetation are very complex, not only because of the great complexity of Mediterranean ecosystems and the interactions with land uses, but also because of the different responses to different type of fires and fire regimes (i.e. different intensities, seasonalities, recurrences and extent of the fire). At the landscape level, post-fire regeneration would depend mainly on the initial vegetation, that is, plant traits of the initial species occurring on the site, and on-site environmental factors (climatic and terrain parameters).

2.4.1
Plant traits

There are two main plant traits conditioning the regeneration pattern: (a) the capacity to resprout after fire (resprouter species), and (b) the stimulation of the recruitment by fire (seeder or recruiter species). We consider fire-stimulated species to be those in which fire stimulates or facilitates the recruitment process (seed dispersal, germination, flowering, etc.) by some physical or chemical mechanism (e.g. heat, smoke). Species that increase after fire because there is more light/space available (opportunistic species) are not considered fire-stimulated species. Species may resprout or not, and may have their recruitment stimulated by fire or not, in four possible combinations (Pausas 1999a): resprouters without recruitment stimulated by fire (R+S-), resprouters with recruitment stimulated by fire (R+S+), non-resprouters with recruitment stimulated by fire (R-S+, obligate seeders) and non-resprouters without recruitment stimulated by fire (R-S-).

These functional types have different demographic patterns and responses to repeated wildfires (for details see Pausas 1999a). Resprouting species always maintain some biomass alive (often below-ground biomass) and recover quickly from fire. A typical example is the *Quercus coccifera* (kermes oak, the dominant species of the garrigues), a vigorous resprouter shrub with rhizomes that quickly recover from fire. Trabaud (1991) experimentally burnt this species every 2 years for 19 years and it kept resprouting. Ferran et al. (1998) have demonstrated the loss of some growth capacity in this species after recurrent fires. This loss can be due to the depletion of carbohydrates and nutrients stored in the below-ground system. Other *Quercus* species have demonstrated their high resprouting capacity from basal buds (e.g. *Q. ilex*). An interesting case is *Quercus suber* (cork-oak) which is able to resprout from stem buds (Pausas 1997) thanks to protection by a thick bark (cork). It is the only European tree species that resprouts from stem buds, as do most Australian *Eucalyptus* (Gill 1981, Strasser et al. 1996), and it produces a quick regeneration of the landscape after fire. This feature, together with its economic importance, makes this species a good candidate for reforestation programs in fire-prone areas.

The recovery of non-resprouting species is slower and depends on the fire interval, the age of maturity (to produce seeds for regeneration) and seed longevity, and resistance to fire. Species with fire-stimulated recruitment show a peak phase soon after a fire, and then they decrease due to their low competition ability. Different recruitment processes may be stimulated by fire, with germination being the most important. Seeds of some species have innate (primary) dormancy and require a fire-related stimulus to germinate (refractory seeds). Fire may also stimulate flowering or dispersal. Examples of species with fire-stimulated germination are most of the Cistaceae and Papilonaceae (legumes) species (e.g. Thanos et al. 1992, Arianoutsou and Thanos 1996). Furthermore, it has been postulated that post-fire vegetation may be rich in legume species because their capacity to fix nitrogen may alleviate nutrient losses caused by fire. However, while some studies seem to support this hypothesis (e.g., in the Greek phryganas, Arianoutsou and Thanos 1996), others do not (in French and Iberian garrigues, Trabaud 1992, Pausas et al. 1999).

Another adaptation to fire is the serotiny (=bradyspory), that is, the retention of the seed in the canopy until a fire occurs (fire-induced seed dispersal). This fire stimulation system is frequent in other Mediterranean type ecosystems (South Africa, Australia) but in the Mediterranean basin it is only found in a relatively low level in a few species such as the Mediterranean pines (e.g. *Pinus halepensis*, *P. brutia*). In these pines the recruitment is stimulated by fire because of increased seed dispersal rather than germination stimulation. In contrast with some other Mediterranean type ecosystems (such as South Africa, Australia), in the Mediterranean Basin no plants have been found to be strictly dependent on fire for completing their life cycle.

There are some species which possess both the capacity to resprout after fire and to have their recruitment stimulated by fire (e.g. *Thymus vulgaris, Anthyllis cytisoides, Dorycnium pentaphyllum*). However, in these cases both factors are usually developed in a lesser degree (i.e. lower resprouting capacity and lower stimulation of the recruitment) than for resprouting (non-stimulated recruitment) species and obligate seeders.

2.4.2
Environmental conditions

Environmental conditions affect both fire behaviour and vegetation response, and the differential effect on each of these factors is difficult to separate from field observations. Several samplings in the eastern Iberian Peninsula have been undertaken to test the effect of different environmental (terrain and climate) parameters on regeneration (Vallejo 1997, Pausas et al. 1999).

Two main bioclimatic zones from the point of view of temperature and two main bedrock types are found in the eastern Iberian Peninsula (Valencia region): the thermo-Mediterranean zone right next to the coast with a mean annual temperature between 17-19 °C (vegetative period = whole year) and the meso-

Mediterranean zone, somewhat inland with a mean annual temperature between 13-17 °C (vegetative period = 9 - 11 months). The two main bedrock types are: limestone, that is, calcareous hard rocks producing very shallow and decarbonated brown-red soils with abundant outcrops and cracks, and marls which produce deeper and highly carbonated soils but without cracks. The combination of these factors (bioclimatic region x bedrock type) gives four distinctive environmental conditions. Several sites were selected in each of these environments after the 1991 fires, and the vegetation covers were analysed 10 and 34 months after the fire (Vallejo 1997, Pausas et al. 1999, Fig. 2.4). Results clearly show the influence of bedrock type, while differences in climatic region were less evident: plant recovery was significatively lower on marls in both samplings (10 and 34 months after fire). Ten months after fire the mean plant cover was over 71% on limestone bedrock type, and ca. 40% on marls. At that time two highly resprouting species (both having rhizomes), *Quercus coccifera* (kermes oak) and *Brachypodium retusum* (a perennial grass) covered ca. 30 and 46% on limestone and 7 and 24% on marls.

The effect of facing slope on post-fire recovery was studied after the 1994 large fire in the meso-Mediterranean bioclimatic zone under limestone bedrock of the eastern Iberian Peninsula. A year after the fire, on the average, vegetation covered 42% of the soil, and the cover was significantly greater on the north slopes (52.4%) than on the south ones (32%, Fig. 2.5). This recovery was mainly due to the resprouting species *Quercus coccifera* and *Brachypodium retusum*, and to the obligate seeders *Ulex parviflorus*, *Helianthenum marifolium* and *Cistus albidus*.

The regeneration of *Pinus halepensis* has also shown some relation with envi-

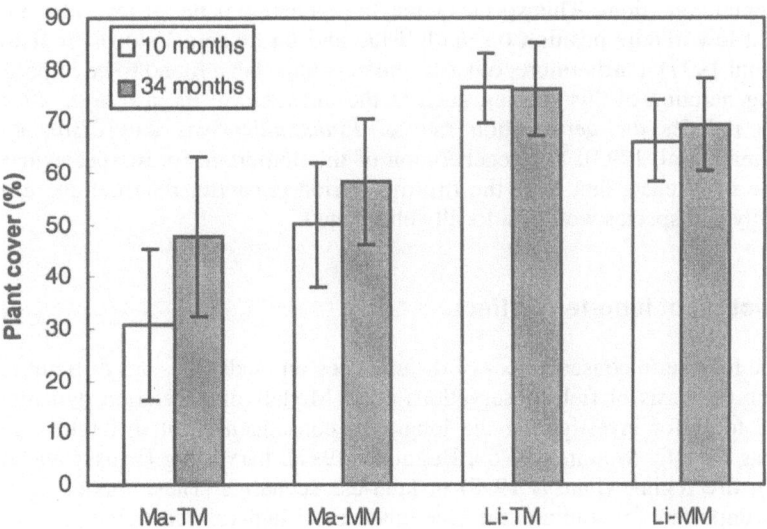

Fig. 2.4. Plant cover (%) 10 and 34 months after fire in four environmental conditions: two bedrock types (Ma: marls, Li: limestones) and two bioclimatic regions (TM: thermo-Mediterranean and MM: meso-Mediterranean). Vertical lines are standard deviations. Elaborated from data in Pausas et al. (1999).

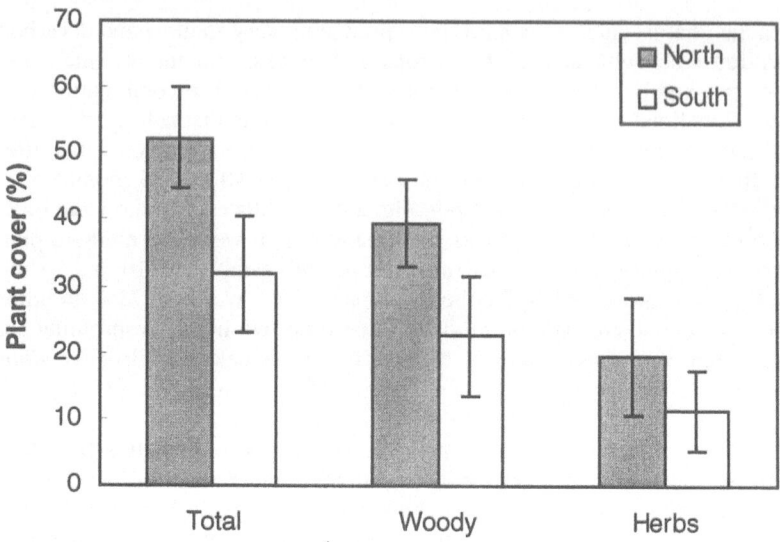

Fig. 2.5. Vegetation cover (%) for total and for two species groups (woody species and herbaceous species), on different facing slopes, one year after the 1994 fire on limestone bedrock type and under meso-Mediterranean bioclimates in the Eastern Iberian Peninsula. Vertical lines are standard deviations. Elaborated from data in Pausas et al. (1999).

ronmental conditions. This species tends to regenerate better at low altitude (Fig. 2.6), at low/middle position on the hillside and on moderate slopes or flat areas (Tsitsoni 1997). Furthermore, other parameters have been found to be important in the regeneration of this species, such as the thickness of the ash layer after fire, which reduces the germination rate of *Pinus halepensis* and *Cistus* species (Ne'eman et al. 1993). The regeneration of this important forest species depends on fire recurrence, that is, if the fire-free period is shorter than the age to reach maturity, the species would be locally eliminated.

2.4.3
Prediction of long-term effects

The long-term consequences of disturbances on landscapes are difficult to predict on the basis of field observations only. Models of vegetation dynamics are useful tools for investigating the long-term consequences of different scenarios such as climatic (Solomon 1986, Bugmann 1997), harvesting (Pausas and Austin 1997), fire regime (Pausas 1998) or land-use scenarios. These models are especially important for studying the consequences of interval-dependent processes (in contrast to event-dependent processes; Bond and van Wilgen 1996), where the experimental approach is difficult to apply. Interval-dependent processes such as establishment, maturation and dormancy are key factors for predicting the long-term consequences of alternative fire scenarios on Mediterranean landscapes.

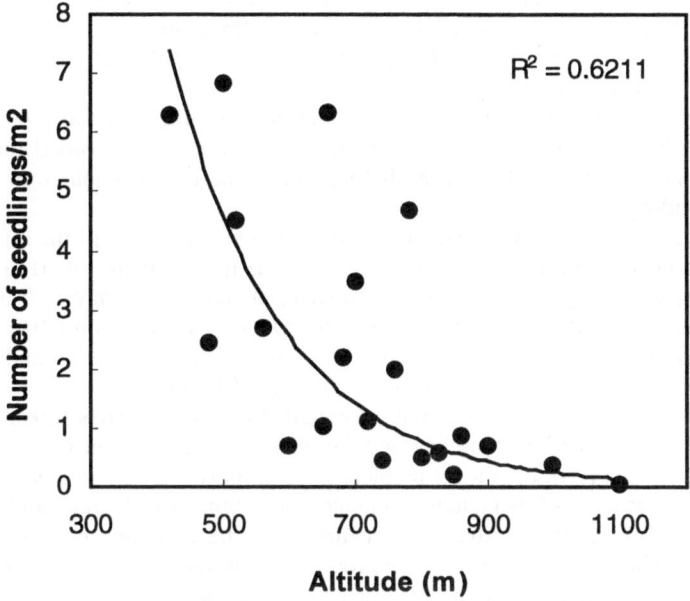

Fig. 2.6. Mean number of pine (*Pinus halepensis*) seedlings per m² under burnt adults along an altitude gradient. Elaborated from data in Pausas et al. (1999).

Modelling vegetation dynamics in fire-prone ecosystems requires two steps (Pausas 1998). The first is to produce descriptive (often statistical, e.g. Strasser et al. 1996, Pausas 1997) models based on field observations after wildfire or experimental fire. These models are a simplified description of the response of vegetation to disturbance events (fires). The second step is to introduce ecological knowledge of how plants function and descriptive models from field observations into a mechanistic simulation model. This simulation model allows us to simulate long-term vegetation dynamics, and to test alternative scenarios or management options.

The most standard modelling technique currently used to predict vegetation dynamics is the gap model approach (individual-based gap dynamic modelling approach, Botkin et al. 1972, Shugart 1984) which simulates stand development by modelling mechanistically the establishment, growth, and death of individual plants. We have developed a simple gap model, called BROLLA (Pausas 1999b), that simulates the establishment, growth, and death of individual plants in a 200 m² plot and at annual time-steps. BROLLA aims at capturing the main features for modelling the dynamics of Mediterranean vegetation. It is based on other gap models (Botkin et al. 1972, Shugart 1984, Coffin and Lauenroth 1990, Pausas et al. 1997) and includes four species groups (plant functional types) growing in the north-eastern Iberian Peninsula: *Quercus* (tree, broad-leaved evergreen resprouter;

e.g. *Q. ilex*), *Pinus* (tree, needle-leaved non-resprouter with serotinous cones; e.g. *P. halepensis*), *Erica* (heath, ericoid-leaved resprouter; e.g. *E. multiflora*), and *Cistus* (shrub, broad-leaved non-resprouter with germination stimulated by fire; e.g. *C. albidus*). Six fire scenarios were simulated for a 500-year period (mnemonic names in brackets): no fire (NF) and fire every 100 (F100), 40 (F40), 20 (F20), 10 (F10) and 5 (F5) years. The objective of these fire scenarios was to create a fire recurrence gradient to study the behaviour of the functional types with changes in fire recurrence. In the presented model runs, it is assumed that the simulated patch is within a large fire, that is, there is no external source of seeds from surroundings.

The BROLLA model predicted changes in the relative abundance of the different species with changes in the fire recurrence in the NE Iberian Peninsula (Fig. 2.7). *Quercus* showed a progressive decrease in relative abundance, from ca. 70% (NF) to less than 5% (F5) of the total basal area. *Pinus* had its maximum relative abundance (30%) at low and intermediate fire recurrences (NF-F20). *Erica* and especially *Cistus* increased with increased fire recurrence (F10 and F5).

Without fire, BROLLA predicts a forest dominated by *Quercus* (oak forest), with some *Pinus* and a very low presence or total absence of *Erica* and *Cistus*. With high fire recurrence (F5), the model predicts a community (shrublands) dominated by *Erica* and *Cistus*, with very low abundance or absence of *Quercus* and *Pinus* (immatures only). At intermediate fire recurrences, the predicted community is dominated by *Quercus* with *Pinus* (mixed forest) depending on the fire recurrence. These results are roughly as expected for eastern Spain (Terradas 1987, Ferran

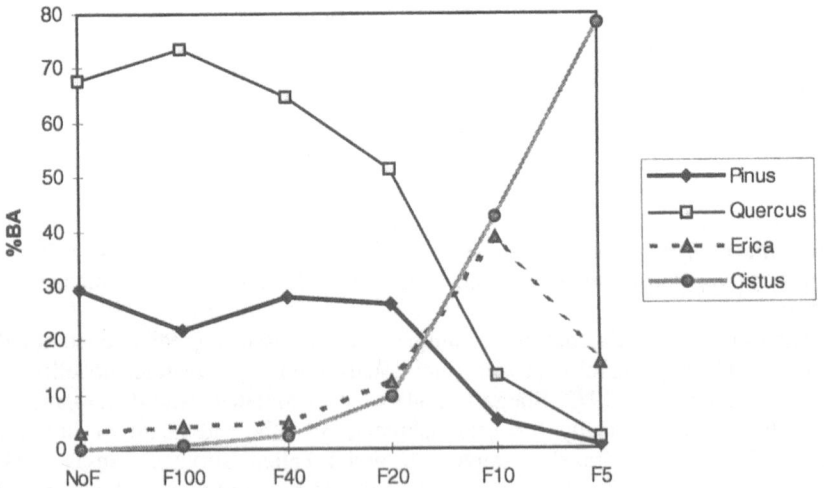

Fig. 2.7. Summary of the BROLLA simulation results for six scenarios with four species groups (named *Quercus*, *Pinus*, *Cistus*, *Erica*). No fire (NF) and fire every 100 (F100), 40 (F40), 20 (F20), 10 (F10), and 5 (F5) years. The y-axis represents the mean percent basal area of ten replicated simulations for 500 years. From Pausas (1999b).

1996), that is, mature oak (e.g. *Q. ilex*) forest in areas with low fire recurrence, pine/oak woodlands (e.g., *P halepensis*) in areas with higher fire recurrence, and shrublands (e.g. *Erica multiflora* and *Cistus* sp. pl.) in the highest fire recurrence areas. Similar patterns of increasing shrublands and decreasing *Quercus* communities due to increasing fire recurrence are found in other Mediterranean basin systems (e.g. Trabaud and Galtié 1996). In a successional study using a chronosequence in the NE Iberian Peninsula (Ferran 1996), *Quercus ilex*, *Pinus halepensis*, and *Cistus salviifolius* followed a similar pattern to the one simulated here, although *Erica* species were less abundant. That is, *Quercus* was abundant in the mature stands (23 and 35 years since last fire), while *Cistus* was very abundant in the young ones (2 years since fire). *Pinus* was low in abundance or absent in all the chronosequence, and absent in the young stands. A *Quercus coccifera* garrigue in eastern Spain (Ferran et al. 1998) showed a significant decrease in *Quercus* biomass from a plot unburned at least for the last 18 years to similar garrigues burnt once, twice, and three times in that period; the cover of *Cistus* species (biomass was not sampled) showed an opposite trend. Model results are consistent with these trends although accurate comparisons are not possible. It is very difficult, if not impossible, to verify accurately these types of models because of the difficulty in obtaining reliable fire history information at the patch level, and because of the spatial variability in the response. Current vegetation is a mosaic of different patches, with different fire histories and different site (e.g. soil, microclimate) characteristics.

The land-use history may interfere with the dynamic pattern described here. Cultivation requires the uprooting of resprouters, and although land abandonment permits new plant colonisation, this is often difficult for many sclerophyllous resprouter species because of their low dispersal capacity and/or their requirement of an animal vector for dispersal (zoochory) which may not be available in degraded lands (Alcántara et al. 1997). In addition, germination rates and seedling survival are often low for these species in degraded drylands, as is the case of *Quercus rotundifolia* (Laguna and Reyna 1990, Pausas 1999a).

Vegetation dynamic models provide us with some insights into the fate of vegetation with changes in fire recurrence, and can be used to test alternative management scenarios. However, further work is needed before we can obtain reliable models for global change predictions. The main research needed includes: the extension to spatially explicit gap models (including realistic dispersion processes), realistic environmental (especially temperature) response of the species, knowledge of key species attributes, and the inclusion of below-ground competition. Research on these topics would improve our ability to produce more accurate simulation models.

2.5
Concluding remarks

Large forest fires have spread at an unprecedented rate in southern Europe during the last decades, after dramatic changes in land use and population distribution between urban and rural areas. Similar trends could be predicted for northern African countries in the future. The challenging questions now are to what extent the present fire regime can be modified through suppression and prevention policies, and what might be the consequences for ecosystems and landscape structure and dynamics. To gain some insight into these questions, ground-based observations should be coupled with remote sensing techniques in order to provide the knowledge basis for modelling various management scenarios at the appropriate scale.

Acknowledgements. This research was supported by the government of Valencia (*Generalitat Valenciana*) and the Spanish government (*Ministerio de Educación y Ciencia*).

3 Short-term fire risk: foliage moisture content estimation from satellite data

Emilio Chuvieco[1], Michel Deshayes[2], Nicholas Stach[2]
David Cocero[1] and David Riaño[1]

[1]Department of Geography, University of Alcalá (Spain).
[2]CEMAGREF - ENGREF Remote Sensing Laboratory (France).

Abstract. A description of methods used to determine short-term changes in fire danger is reviewed, mainly those based on the estimation of foliage moisture content (FMC). Applications of low-resolution data, acquired by the Advanced Very High Resolution Radiometer (AVHRR) on board the NOAA satellites, as well those based on high-resolution data, are examined. Examples of FMC estimation on Mediterranean areas are also presented, both for AVHRR and Landsat-TM data. In both cases, satellite data provide a higher confidence to estimate FMC in grasslands than in shrublands, although in both cases, some variables provide significant correlation, especially when the spring season is taken into account. The most sensitive variables for FMC estimation are based on short-wave infrared bands, and the combination of vegetation indices and surface temperature.

3.1
The role of foliage moisture content in the short-term estimation of fire danger

The estimation of fire danger may be considered on different spatial and temporal scales. This chapter is focused on the contribution of satellite remote sensing data to estimate fire danger regarding the most dynamic factors of fire ignition or fire behaviour, while Chapter 5 deals with long-term trends of fire danger.

As is well known, forest fire occurrence is a result of three main factors: a heat source, a fuel that burns, and oxygen (wind) to propagate combustion. Ignition sources may be natural or human-caused, the latter being the most common in the Mediterranean areas. Fuel refers to the different components of the vegetation layer, both live and dead materials. Atmospheric conditions influence both fire ignition and propagation, since they increases vegetation dryness and provides oxygen for fire propagation.

Obviously, the contribution of remote sensing to fire danger estimation should be centred on determining fuel composition and state. The former addresses the question of a better understanding of factors related to vegetation structure, which in turn affect fire propagation. These are mainly the compactness, volume to surface ratio, vertical and horizontal continuity, and particle size. Since these variables may present a great variety even in the same vegetation community, fire scientists have created fuel models, which summarise vegetation structural properties in a few types, which can be more easily identified and mapped. The best known fuel types are those produced for the American NFDRS (Bradshaw et al. 1983; Deeming et al. 1977), the Canadian CCFRS (Alexander and McAlpine 1987) and the Behave fire behaviour program (Andrews 1986; Burgan and Rothermel 1984). The contribution of remote sensing data to fuel type mapping has been mainly concentrated on high-resolution sensors or aerial photography, although some global projects have also been developed (see Chap. 5).

Another critical parameter of fuel properties related to fire danger refers to its water content. The amount of water in the fuels has important effects on both fire ignition and fire behaviour. In this respect, fuels refer not only to live plants, but also to the dead materials lying on the forest floor. The drying of any of these two components (live or dead) increases ignition probability and fire rate of spread. Consequently, the estimation of fuel moisture content is taken into account in most fire behaviour models (Albini 1985, Andrews and Chase 1990, Hartford and Rothermal 1991, van Wagner 1985). The contribution of satellite data to the estimation of moisture content relies on a better understanding of the effects of the amount of water on plant reflectance or emittance. The challenge here is to prove that changes in reflectance or emittance produced by water content variations are distinguishable from other factors, such as soil background reflectance and the effects introduced by leaf geometry and observation angles. This chapter presents a critical review of what has been done in this direction, and what future sensors could offer to this research objective. Since satellite-measured reflectance is much more related to leaves than to the wooden parts of the plant, the estimation of fuel moisture content is basically related to the foliage conditions, which on the other hand are more sensitive to assess the water status of the plant. Consequently, the term foliage moisture content (FMC) will be used throughout this chapter.

3.2
The estimation of foliage moisture content

In spite of the importance of FMC in fire danger modelling, few operational methods have been proposed to estimate this variable. The most reliable are field sampling and meteorological moisture codes, but both present shortcomings.

Undoubtedly, the trustiest method to estimate FMC is based on field sampling, since a direct measurement on plant water status is directly performed. However, this procedure is costly and does not assure spatial significance. Although FMC refers to the whole plant, moisture content is commonly measured only from the leaves, at least in live fuels, since wooden parts of the plant are less sensitive to atmospheric variations. Several measures of FMC have been proposed (Desbois et al. 1997a), the most common being the proportion of wet over dry weight:

$$FMC = \left(\frac{W_w - W_d}{W_d} \right) * 100, \tag{1}$$

where W_w is wet weight and W_d dry weight of the same sample. Other measurements of plant water content (such as the Relative Water Content, RWC, that is a function of actual versus potential maximum moisture content) are more common than FMC in plant physiology, for instance to determine plant water stress (Grace 1983). However, FMC is most common in fire-related studies (Brown et al. 1989, Viegas et al. 1990), since the actual amount of water is the critical parameter in fire behaviour. Different spatial sampling techniques have been proposed for FMC measurements, the most typical being transects and quadrats.

Samples are commonly composed of leaves for trees and shrubs, while for grasslands the whole plant is measured (Desbois et al. 1997a). Samples are weighed on the field with portable balances or introduced in ice bags to be weighed at the laboratory. Then, these samples are dried in an oven and weighed again to compute FMC. The length and temperature of the drying process varies among the authors. Some suggest 24 or 48 hours at 60 °C, while some others propose 24 hours at 100 °C. The former is more secure to avoid loss of oils and essentials, while the latter is less tedious (Viegas et al. 1998).

The estimation of FMC from meteorological indices is the most standard method used by national forest services and fire behaviour models. In this case, an indirect estimation of FMC is pursued, assuming that FMC can be predicted following atmospheric trends. However, not many studies have actually verified this relationship, which may greatly vary among the species depending on their physiological properties (especially root length and adaptation to summer drought). For this reason, most meteorological moisture codes available are adapted to dead fuels (Deeming et al. 1974, Cheney 1968), which are more clearly dependent on atmospheric variability than live plants. On the other hand, dead fuels are the most easily flammable. Although some studies have been carried out on the relation between some meteorological codes and FMC (Viegas et al. 1998), further research should be devoted to establish more general conclusions.

In any case, these indices are computed for the locations where meteorological stations are available. Therefore, an additional source of error is introduced to spatially interpolate these indices. In many cases, weather stations may be quite far from fire-prone areas. Consequently, new alternatives to derive FMC information would be greatly appreciated by fire managers.

3.3
The effect of moisture content on reflectance and temperature

Several laboratory and field analyses have been conducted in the last years to improve our knowledge on the effects of water content on vegetation reflectance and temperature, which is the basis for applying remote sensing methods to FMC estimation.

Works focused on optical bands have tried to estimate water content from variations in leaf reflectance, either from raw spectral data or from synthetic indices (ratios, tasseled cap, etc.). Thermal studies have related leaf water content to evapotranspiration (ET) rates, which can be detected by the cooling effect produced by the latent heat loss.

Optical reflectance has been related to water content especially in the short-wave infrared region (SWIR), between 1.4 and 2.5 µm (Bowman 1989, Cohen 1991, Hunt et al. 1987, Jackson and Ezra 1985, Ripple 1986, Thomas et al. 1971, Tucker 1980). Since these wavelengths present the highest absorption due to water, SWIR reflectance is negatively related to leaf water content. In other words, the lower the reflectance, the higher the FMC, since water absorbs strongly in this spectral band.

Regarding near-infrared (NIR) reflectance (0.8 - 1.1 µm), conclusions gathered by different studies do not completely agree. Some authors find a small increment as the leaf dries, which may be caused by the increase in the refractive index of the mesophyll layer when water is replaced by air (Bowman 1989; Hunt and Rock 1989; Hunt et al. 1987; Thomas et al. 1971). Other authors do not find a significant change in NIR reflectance for leaf drying (Carter 1991), while some others measure a clear decrease in reflectance (Westman and Price 1988). This reduction may be caused by the indirect effects of leaf dryness, such as the shrinkage in leaf area index or shadowing due to leaf curling (Jackson and Ezra 1985, Westman and Price 1988).

Finally, the visible domain does not present a great sensitivity to estimate leaf water content, at least in absolute terms. On one hand, the chlorophyll reduction after drying increases reflectance, while on the other the decline in water decreases reflectance. Some authors, though, found that the red band is sensitive to water content (Jackson and Ezra 1985; Ripple 1986), while some others did not observe significant changes (Bowman 1989, Thomas et al. 1971).

When these laboratory studies are applied to the use of remote sensing images for FMC estimation, models to extrapolate from leaf to canopy reflectance are required. In this regard, some authors emphasise the difficulties in estimating plant water content from remote sensing measurements, since they consider that the contribution of leaf water is too weak with respect to other factors affecting plant reflectance: leaf geometry, shadows, soils, etc. (Carter 1991, Cohen 1991). However, other authors have found high correlation between leaf water content and canopy reflectance (Cibula et al. 1992, Hunt and Rock 1989, Hunt et al. 1987, Westman and Price 1988).

An alternative to determine FMC is the analysis of plant thermal dynamism. Rises in air temperature will also increase evapotranspiration rates when water is available to the plant. This increase in latent heat reduces sensible heat, and consequently decreases leaf temperature. On the contrary, when the plant dries, transpiration is reduced and, consequently, so does latent heat, whereas sensible heat increases simultaneously (Kozlowski et al. 1991). As a result of this relation, the difference between air and surface temperature should be clearly related to plant water content and to water stress. Several indices have been proposed based on this difference and on other weather variables (net radiation, air humidity, wind data, etc.). The most currently used are the Stress Degree Day (SDD) (Jackson 1986), the Crop Water Stress index (CWSI) (Jackson et al. 1981), the Water Stress Index (WSI) (Vidal et al. 1994) and the Water Deficit Index (WDI) (Moran et al. 1994). The WSI and WDI have been successfully tested as a predictor of fire danger with, respectively, NOAA-AVHRR and Landsat-TM data (Vidal et al. 1994, Vidal and Devaux-Ros 1995).

3.4
The use of low resolution data for foliage moisture estimation

The estimation of FMC from satellite data has been attempted both with high and low resolution sensors. The former reduce one factor of noise in quantitative correlation with field data, since they provide better accuracy in locating the land plots (Vidal and Devaux-Ros 1995). The latter offer a higher temporal resolution, and therefore are more likely to be used operationally, since forest managers require frequent updates of FMC.

The most commonly used sensor in FMC estimation has been the Advanced Very High Resolution Radiometer (AVHRR) on board the NOAA satellite series. This sensor was initially launched in 1979, and since then it has been included in the different missions of NOAA satellites (Cracknell 1997). Currently, it is operational on the NOAA-12, 14 and 15 satellites. AVHRR images are acquired twice a day, at a 12-hour interval. The NOAA-12 and 15 acquire AVHRR images over the Mediterranean at 8:30 and 20:30 GMT (approximately), while NOAA-14 operates at 2:30 and 14:30 GMT. The latter is the most widely used in FMC estimation, since it is closer to the period of maximum temperature and therefore observes the most dangerous conditions.

Direct estimation of FMC from AVHRR data has been attempted with grasslands in Australia (Chladil and Nunez 1995, Paltridge and Barber 1988, Paltridge and Mitchell 1990) and the USA (Eidenshink et al. 1990, Hartford and Burgan 1994). Grasslands are more efficiently observed than other fuels by remote sensing methods, since their surface is more homogeneous and is mostly covered by leaves, with very little or no proportion of woody materials, which are less sensitive to atmospheric variations. Additionally, grasslands are in many cases annual plants and have less root length. Therefore, they are more dependent on seasonal conditions than shrubs or trees. The referred works have found good agreements between FMC of grassland and AVHRR multitemporal series. Images from the spring and the summer season have been used to consider a wider range of greenness conditions. Correlation with vegetation indices (particularly the Normalised Difference Vegetation Index, NDVI) was generally found to be significant, with positive trends (the higher the NDVI, the higher the FMC). Experiences with shrubs have been less successful, with diverse trends regarding the different species analysed. In a pilot work developed in the South of Spain good correlations were found for *Cistus ladanifer* and *Rosmarinus officinalis,* two very common Mediterranean species, while *Erica australis* offered poorer trends (Alonso et al. 1996). In this case, a ratio of NDVI and Surface Temperature (ST) provided the best results.

Additional studies have also suggested the use of AVHRR data to monitor fire danger (Desbois et al. 1997b). Observed fire occurrence is put in relation with multitemporal trends of AVHRR images acquired before the fire, assuming they contain information on vegetation dryness, an important factor in fire danger (González et al. 1997, Illera et al. 1996, López et al. 1991, Prosper-Laget et al. 1994, Prosper-Laget et al. 1995, Vidal et al. 1994). These studies propose different techniques to emphasise the multitemporal decrease in vegetation vigour, mainly by using indices based on the measure of change from previous periods (González et al. 1997, Illera et al. 1996, López et al. 1991). An assessment of their methods is performed by comparing changes in NDVI or ST with fire occurrence, but actual field estimations of FMC are not performed. This approach may be considered an indirect validation of their methods, since fire occurrence is not always related to FMC. Fire danger is a conjunction of different factors, both physical and human-caused. Satellite data can only assess vegetation dryness (or, to be precise, moisture content), but other factors related to fire ignition or fire propagation cannot be directly derived from satellite observations. Fire only occurs when an ignition cause is present, even if FMC is not critically low. On the other hand, critical levels of FMC may not necessarily lead to fires, if other factors of risk do not appear.

A very sound approach to integrated analysis of fire danger is based on the combination of satellite data and meteorological danger indices. The former would inform on live fuel conditions, while the latter would provide an estimation of FMC for dead fuels. Theoretical frameworks are available (Burgan et al. 1998), but additional research is required to obtain a proper integration of these two sources of information (Aguado et al. 1998, Burgan and Hartford 1993, Burgan et al. 1998).

3.5.
Application of NOAA-AVHRR data to FMC estimation

3.5.1
Study areas

Within the framework of the Megafires project, (FMC) measurements have been carried out in 1996 and 1997 in three study areas (Cabañeros, Spain; Les Maures, France and Chalkidiki, Greece: Fig. 3.1) and compared with NOAA-AVHRR-derived indices. An additional set of FMC measurements performed in France by the French National Forestry Board (ONF) has been made available and has been added to the protocol.

The Spanish test site, Cabañeros National Park, covers 40,000 ha and is located in central Spain, in the autonomous community of Castilla-la Mancha. It is composed of three main units of vegetation: forest of *Quercus suber, Q. rotundifolia* and *Q. faginea*, shrublands with *Arbutus unedo, Rosmarinus officinalis, Phillyrea angustifolia, Pistacia terebinthus, Erica australis, E. umbellata* and *Cistus ladanifer,* and grasslands. Elevation ranges from 500 to 1043 m. The sampling scheme developed for the Megafires project comprised three plots of grasslands, two of shrublands and one plot with *Q. faginea* (in 1997 only). The FMC measurements were carried out every 8 days from April to September, according to the protocol designed for the project (Desbois et al. 1997a).

The Greek test site is situated in the Prefecture of Chalkidiki (North Central part of Greece). In this prefecture, and more precisely on the Sithonia peninsula, 5 test sites were selected for foliage moisture measurements with an 8-day periodicity. Three test-land plots (A, B and C), with slopes varying from 10% to 60%, are dominated by *Quercus coccifera*, alone or with *Q. ilex,* totalling from 40% to 60%

Fig. 3.1. General location of the study areas.

of the cover, the rest being composed of other species like *Arbutus, Erica* or *Juniperus*. The two other test-land plots have gentle slopes and are completely covered with one shrub species, *Cistus* sp.

The first French test-land plot is the forest massif of Les Maures, which is located in the *Département du Var* (southeastern France) and covers almost 90,000 ha. Elevation ranges from 0 to 780 m and the topography is rough. Well individualised and rather homogeneous, this area is mainly covered by cork oak (*Quercus suber*), in association with holm oak (*Q. ilex*) and pines (*Pinus halepensis, P. pinaster*) and various types of shrubs, especially *Erica arborea* and *Arbutus unedo*. In this massif, the INRA, the French Agronomic Research Institute, has established an experimental station, where *Erica* and *Arbutus* FMC (and inflammability) have been monitored every summer since 1989 (Valette 1990). In order to study the spatial variations of FMC, from 1995 onwards, an additional network of 13 plots was established in shrublands all over the massif, where the FMC of the same two species have been measured twice a week. On the plot situated near the experimental station, the FMC of *Erica* is measured five times a week, instead of two (Moro 1996).

Additionally, a second set, or network, of French test-land plots has been used. This network, funded by the *Délégation à la Protection de la Forêt Méditerranéenne*, is operated by ONF. It covers the whole French Mediterranean zone, comprising 15 *départements*. The aim of this network is, as an operational support to fire fighting authorities, to monitor the evolution of foliage moisture on a regional (85,000 km^2) and *département* scale (i.e. for an area of about 6,000 km^2). The experiment is based on the INRA experience on Les Maures, but concerns a greater number of species and a larger area. Thirty measurement points were chosen, located within a 10-km radius from a meteorological station, within a homogeneous forest massif representative of the *département*. For each point, two shrub species, typical and dominant in the area, were selected and sampled. The more represented species, at least on two land plots, are one tree (*Quercus ilex*) and different shrubs (*Cistus monspelliensis, C. albidus, Erica arborea, Rosmarinus officinalis; Genista cinerea, Quercus coccifera* and *Juniperus oxycedrus*). FMC measurements are performed on these species twice a week, on Mondays and Thursdays, from July to September. Thirty six hours after collection, the FMC data are supplied to the local and regional fire fighting authorities.

3.5.2
Satellite data processing

AVHRR images were acquired daily from the afternoon pass of NOAA-14 satellite. The study period comprised from 1 June to 30 September 1996 and 1997, between 12:30 to 15:30 (GMT), which is the time of maximum vegetation dryness. Additional images were purchased over the Iberian Peninsula to cover two spring months (April and May) in which FMC samples were available.

The pre-processing chain applied on the images, detailed in Vidal et al. (1997), consisted of the following steps:
- calibrating visible and near infrared bands (channels 1 and 2) to convert the raw digital counts into reflectance at the top of atmosphere,
- splitting the middle infrared band's (channel 3) total radiance into its emitted and reflected components,
- converting the digital counts of thermal infrared bands (channels 4 and 5) into radiance and then from radiance to surface temperature,
- correcting the geometry of the images using the SGP4 orbital model.

After preprocessing, the images have 8 channels: VIS, NIR, MIR reflective component, MIR emitted component, T4, T5, ST, NDVI. In 1997, a view angle channel was added. These channels have been used to compute several indices mentioned in previous fire danger studies. The main steps of the processing, detailed in Desbois et al. (1997b), were the following:
1. Image subset and manual registration: after standard registration through the SGP4 orbital model, the residual position error of the images reached 3 or 4 pixels, which was not satisfactory. Therefore, a subset of each daily image was extracted over each study area and a manual geometric shift has then been performed to register it with coast lines (or ground control points when available).
2. Selection of good quality images. Different defects affected the preprocessed images: cloud cover, high incidence angle, bad junction in mosaics, missing data. A visual examination of the images and of the NDVI and ST temporal profiles of each study land plot was carried out to detect and eliminate unsuitable images.
3. Generation of maximum value composite images. Besides daily images, Maximum Value Composites (MVC) images were generated. Though somewhat too long for an operational application, and slightly below the 9-day revisit periodicity of AVHRR, but in accordance with the FMC sampling periodicity, a composite period of 8 days for Greece and Spain, and 7 days in France was chosen. The MVC was processed by assigning to each pixel: (a) the maximum NDVI of this pixel for the 8-day (or 7-day) period, and (b) the ST of the day where the NDVI was maximum. For France, due to the multi-weekly rhythm of measurements, different independent sets of MVC were prepared, one for each day of measurement per week, with composite periods ending on that day.

4. Indices calculation. Three different types of indices were tested, related to vegetation indices, to surface temperature or to a combination of both. The first type included several vegetation indices (NDVI, Rouse et al. 1974, SAVI, Huete 1988, and GEMI, Pinty and Verstraete 1992), as well as indices derived from multitemporal series of NDVI —hereafter named multitemporal NDVI indices: GRN_{rel}, GRN_{abs} and SM, Eidenshink et al. 1990, and cumulated decrements of NDVI (ARND, López et al. 1991). A second category was based on the thermal behaviour of the vegetation as an indicator of its water stress. Thermal infrared data are then directly used (ST) or combined with meteorological data - especially air temperature - to estimate vegetation water stress (surface minus air temperature ST-AT, water stress index WSI, Vidal et al. 1994). A third category is based on the combination of vegetation indices and thermal data (NDVI/ST, Alonso et al. 1996).

5. Validation of satellite indices with ground data. Fire danger indices derived from NOAA-AVHRR were computed on 3x3 pixel areas centred on each sampling land plot (to minimise the consequences of residual misregistrations) and then correlated with the FMC measured on the same day (for daily computed indices), or measured at the end of the 8-day (or 7-day) period (for indices computed with temporal composite values).

The results are presented in the following section. For presentation clarity the subsequent typographic conventions have been used in Tables 3.1 to 3.8:

- *numbers in italics* (ex. *0.62*) represent correlation with sign opposite to expected,
- numbers in plain font (ex. 0.62) represent correlation with the expected sign,
- **bold numbers** (ex. **0.62**) represent correlation significant at a 95% level of confidence,
- <u>underscored</u> numbers (ex. <u>0.62</u> or <u>**0.62**</u>) indicate the index which is the highest in a line, i.e. the best performer for the (group of) land plot(s) of that line.

3.5.3
Results on the Chalkidiki study area

The correlation between foliage moisture and satellite data on Chalkidiki study area in 1996 are shown in Table 3.1 (daily data) and Table 3.2 (MVC data). First a general remark, valid for all validation land plots, has to be made in the interest of greenness indices, GRN_{rel}, GRN_{abs} and SM. GRN_{abs} is a linear combination of NDVI with fixed coefficients, and therefore presents correlation coefficients which are exactly the same as with NDVI. This index is of great interest to compare among land plots (spatial correlation), but this is not the object of this study. Therefore, results from this index have been dropped from the tables. GRN_{rel} is a linear combination of NDVI too, but with coefficients depending on the land plot: for an individual land plot it gives the same results as the NDVI and only for a group of land plots is it therefore worth being computed. The same is valid for SM, linear combination of GRN_{rel} and GRN_{abs}.

For daily data (Table 3.1), all the correlation values offer the expected sign (but 2 out of 35): correlation of FMC is positive with NDVI and derived indices, and negative with ST. Furthermore, these correlation coefficients are statistically significant for two of the five land plots, land plots D and E, which present a shrubby and homogeneous cover, with *Cistus* only. The global results (all land plots together) are also significant for all satellite indices. Regarding the different types of indices, NDVI-derived indices behave more coherently than the ST index. The correlation with maximum value composite (Table 3.2) are similar to those obtained with daily data. The results are even generally better with the less homogeneous land plots (land plots A to C) two of which (land plots A and B) now show a significant correlation with at least one index. The ST index is still the less homogeneous index, with one positive (instead of negative) correlation.

Table 3.1. Pearson correlation coefficients between FMC and AVHRR-derived indices on Chalkidiki (Year 1996, daily data)

Land plot	Sample size	NDVI	GRNrel	SM	SAVI	GEMI	ST	NDVI/ST
A	6	<u>0.50</u>	0.50	0.50	0.47	0.39	*0.63*	0.07
B	8	0.07	0.07	0.07	<u>0.39</u>	0.31	*0.02*	0.07
C	7	0.17	0.17	0.17	0.19	0.26	<u>**-0.54**</u>	**0.49**
D	8	**0.89**	**0.89**	**0.89**	**0.90**	**0.84**	**-0.58**	<u>**0.91**</u>
E	8	**0.84**	**0.84**	**0.84**	**0.82**	**0.79**	**-0.65**	<u>**0.88**</u>
All	37	**0.44**	**0.36**	**0.41**	<u>**0.55**</u>	**0.43**	**-0.43**	**0.49**

See typographic convention at the end of § 3.5.2

Table 3.2. Pearson correlation coefficients between FMC and AVHRR-derived indices on Chalkidiki (Year 1996, Maximum Value Composite data)

Land plot	Sample size	NDVI	GRNrel	SM	ST	NDVI/ST
A	10	0.54	-	-	**-0.66**	<u>**0.76**</u>
B	10	<u>**0.73**</u>	-	-	*0.19*	0.56
C	10	0.45	-	-	-0.26	<u>0.47</u>
D	10	**0.75**	-	-	**-0.69**	<u>**0.84**</u>
E	10	**0.90**	-	-	**-0.88**	<u>**0.97**</u>
All plots	50	<u>**0.67**</u>	**0.54**	**0.61**	**-0.59**	**0.65**

See typographic conventions at the end of § 3.5.2.

3.5.4
Results on the Cabañeros study area

In the 1996 data analysis (Table 3.3), expected correlation coefficients are confirmed for ST, ST-AT and NDVI/ST indices. ST shows a negative correlation with all species, although *Erica australis* and *Arbutus unedo* offer low values. Similar trends were found with surface minus air temperature (ST-AT). The ratio of NDVI to ST presents the highest correlation among all the satellite variables for grasslands, but ST presents better relations in shrublands.

Regarding vegetation indices, they present a positive sign in their correlation with grasslands, but a negative with shrubs. This may be caused by a shading effect, because while grasslands were sampled in very flat terrain, shrub plots were situated in a rougher area. As a consequence, these plots presented lower NDVI values in the spring season, despite the fact that NDVI is supposed to attenuate shading effects. Apparently, ST and ST-related indices are not so severely affected by shading. The two other vegetation indices show the same trend. The SAVI and GEMI perform slightly better than the NDVI in grasslands, but even worse in shrublands (*Cistus ladanifer* and *Rosmarinus officinalis*), in which they show highly negative correlation (the opposite sign to what should be expected).

Data from 1997 offer a more logical trend with all indices (Table 3.4), every species presenting the expected sign in the correlation. NDVI/ST performs slightly better than ST, offering the best correlation with FMC among all the satellite indices, with all correlation coefficients being significant, and with grassland having the highest r-values. Vegetation indices present now the expected sign in the correlation. It should be noted that the plots of shrub were moved in 1997 to a flatter terrain, less affected by changes in solar illumination. Among the different vegetation indices, the SAVI performs slightly better than the NDVI with most species, the GEMI offering an intermediate value.

Table 3.3. Pearson correlation coefficients between FMC (dry basis) and AVHRR-derived indices in Cabañeros (Year 1996, Maximum Value Composite data)

Site/Species	Sample size	NDVI	SAVI	GEMI	ST	ST-AT	NDVI/ST
Grassland1	21	**0.838**	**0.881**	**0.853**	**-0.780**	**-0.744**	<u>**0.929**</u>
Grassland2	21	**0.760**	**0.844**	**0.805**	**-0.796**	**-0.707**	<u>0.905</u>
Grassland3	21	0.592	0.549	0.314	**-0.754**	**-0.683**	<u>0.870</u>
Rosmarinus officinalis	21	*-0.494*	*-0.715*	*-0.730*	**-0.723**	<u>**-0.751**</u>	0.499
Cistus ladanifer 1	21	*-0.431*	*-0.739*	*-0.763*	**-0.728**	<u>**-0.769**</u>	0.543
Cistus ladanifer 2	21	*-0.132*	*-0.436*	*-0.521*	**-0.704**	<u>**-0.694**</u>	0.639
Erica australis	21	0.079	*-0.015*	*-0.082*	<u>**-0.201**</u>	-0.050	0.180
Arbutus unedo	21	<u>0.199</u>	*-0.085*	*-0.175*	-0.068	-0.154	0.087

See typographic convention at the end of § 3.5.2.

Table 3.4. Pearson correlation coefficients between FMC (dry basis) and AVHRR-derived indices in Cabañeros (Year 1997, Maximum value composite data)

Land plot / Species	Sample size	NDVI	SAVI	GEMI	ST	ST-AT	NDVI/ST
Grassland1	22	**0.713**	**0.765**	**0.784**	**-0.765**	**-0.694**	<u>**0.889**</u>
Grassland2	22	**0.589**	**0.722**	**0.722**	**-0.768**	**-0.572**	<u>0.776</u>
Grassland3	22	0.406	**0.521**	**0.484**	<u>**-0.762**</u>	**-0.643**	0.725
Rosmarinus officinalis	22	**0.423**	**0.451**	0.268	**-0.475**	-0.403	<u>0.492</u>
Cistus ladanifer 1	22	**0.498**	<u>**0.628**</u>	0.412	**-0.594**	**-0.676**	0.624
Erica umbelata	22	0.360	0.279	0.009	-0.409	<u>**-0.580**</u>	0.464
Cistus ladanifer 2	22	**0.703**	<u>**0.684**</u>	**0.522**	**-0.615**	**-0.581**	0.678
Erica australis	22	**0.645**	<u>**0.810**</u>	**0.738**	**-0.580**	**-0.588**	0.658
Quercus faginea	22	0.388	0.269	0.001	<u>**-0.603**</u>	**-0.590**	<u>0.603</u>

See typographic convention at the end of § 3.5.2.

3.5.5
Results on the Les Maures study area

On the Les Maures test site, only the results obtained in 1996 will be presented, as 1997 presents a reduced availability of field and satellite data due a very dry spring (spring shoots missing on dry land plots) and a very humid August (cloud cover and rains). Correlation of FMC with daily data will be presented first, followed by correlation with MVC data.

The correlation between foliage moisture of *Erica arborea* and AVHRR-derived indices for daily data of 1996 are shown in Table 3.5. The fourth column of this table shows the correlation between FMCs of *Erica arborea* and *Arbutus unedo*. The correlation of the latter is rather high, which can be first interpreted as an indication of the local coherence of these FMC measurements. Secondly, this good correlation between the FMCs of the two species explains why the correlations of AVHRR-derived indices with *Arbutus* are very similar to the ones obtained with *Erica*, and are therefore not presented here.

A second remark on Table 3.5 has to be made on the sample size. Whereas the maximum possible size for an individual land plot could be around 24 (3 months x 4 weeks x 2 measurements per week), the sample sizes observed vary between 5 to 10. Two reasons are found for this: first, many images had to be eliminated due to cloud cover, or a too oblique view angle; and secondly, FMC measurements falling on days following rain when foliage was wet had to be cancelled.

The results will be analysed along two axes: according to land plot, or group of land plots, or according to satellite index. As regards analysis by land plot, in Table 3.5, the results are first presented land plot by land plot, then with a partial regrouping by moisture level, following INRA classification (Moro 1996), with fresh (7 land plots), dry (5 land plots) and an intermediate moisture land plot (land plot 6), and a final regrouping with all land plots together (13 land plots).

The important fact to notice is the difference in the significance of the correlation obtained with individual land plots and groups of land plots: generally a land plot does not present any significant correlation coefficient (except with the ARND index, and land plots 3 and 10), but groups of land plots (almost) always have several significant correlation coefficients. The effect seems due not to the fact that the correlation coefficient increases, but that, staying at the same level, it becomes significant because the increase in the sample size lowers the threshold of significance. This is also the case with *Erica arborea* on land plot 5, near the INRA experimental station, with a higher sample size due to its particular 5 times per week sampling scheme.

Table 3.5. Pearson correlation coefficients between *Erica arborea* fuel moisture (dry basis) and AVHRR-derived indices on Les Maures (Year 1996, Daily data)

Plot type	Plot #	Sample size	FMC-AU	Correlation with NOAA AVHRR-derived indices								
				NDVI	GRN rel	SM	SAVI	GEMI	ARND	ST	NDVI /ST	WSI
Fresh sites	1	10	**0.96**	0.46	-	-	0.52	0.47	<u>**0.66**</u>	-0.01	0.29	-0.21
	2	10	**0.87**	0.55	-	-	**0.66**	0.61	<u>**0.72**</u>	*0.17*	0.29	-0.20
	5	28*	**0.85**	**0.49**	-	-	**0.49**	**0.40**	<u>**0.61**</u>	-0.15	**0.39**	-0.16
	7	7	**0.91**	0.42	-	-	0.39	0.32	<u>0.55</u>	*0.37*	0.07	-0.27
	8	7	**0.94**	0.57	-	-	0.28	0.16	<u>0.67</u>	*0.40*	*-0.02*	-0.04
	10	5	**1.00**	0.55	-	-	0.41	0.27	<u>0.70</u>	-0.03	0.37	-0.67
	12	8	**0.99**	0.39	-	-	0.60	0.54	<u>0.66</u>	-0.02	0.22	-0.21
	All fresh	75	**0.90**	0.45	0.30	0.35	0.39	0.29	<u>**0.52**</u>	0.00	**0.26**	**-0.20**
Int.	6	5	0.74	0.77	-	-	0.63	0.46	<u>**0.88**</u>	-0.08	0.54	-0.63
Dry sites	3	9	**0.94**	**0.77**	-	-	**0.72**	0.57	<u>**0.87**</u>	-0.20	0.61	-0.35
	9	5	**0.98**	0.75	-	-	0.69	0.53	<u>**0.87**</u>	-0.01	0.51	-0.51
	11	8	**0.97**	0.52	-	-	0.62	0.49	<u>**0.68**</u>	-0.42	0.54	-0.59
	13	8	**0.96**	0.61	-	-	0.66	0.55	<u>**0.73**</u>	-0.23	0.50	-0.32
	15	10	**0.80**	0.48	-	-	0.23	0.12	0.36	-0.64	0.62	<u>**-0.68**</u>
	All dry	40	**0.91**	<u>**0.62**</u>	**0.59**	**0.61**	**0.56**	**0.42**	**0.59**	**-0.41**	**0.58**	**-0.54**
All plots		120	**0.90**	<u>**0.41**</u>	**0.36**	**0.38**	**0.39**	**0.30**	**0.29**	0.11	**0.31**	**-0.25**

* Site 5 = Les Maures experimental station, 5 measurements per week on *Erica*
Int.: Intermediate moisture site; *FMC-AU*: Correl. with FMC of *Arbutus unedo*.
See typographic convention at the end of § 3.5.2.

When analysing the performance of indices, the best performer, way ahead of the others (in 12 out of 13 individual land plots), is the cumulated relative NDVI decrement, ARND. It is always behaving as expected, showing a positive correlation with FMC, and contrary to others, its correlation coefficient is very often significant, even with small sample sizes. A group of second-best performers comprises other vegetation indices, NDVI, multitemporal NDVI indices (GRNrel, GRNabs and SM) and SAVI, always positively correlated with FMC. On the lower side, the index that performs worst is ST. In about one third of the cases, it shows a positive correlation with FMC, whereas a negative correlation is expected. Another thermal-based index, WSI, behaves better than ST, with correlation coefficients of the appropriate sign, but only a few are significant. The other indices, GEMI and NDVI/ST, show an intermediate efficiency, better than ST and WSI but less efficient than ARND, and even NDVI or multi-temporal-NDVI indices.

Regarding trends between FMC and MVC data on Les Maures, only correlation with *Erica* are presented here (Tables 3.6 and 3.7), the correlation with *Arbutus* following the same pattern. The tables present the results obtained with one of the two series of MVC data, the one with compositing periods ending on Thursdays, where the correlation with the other series presents only minor differences and variations. As with correlation with daily data (Table 3.5), correlation with MVC data for group of land plots (Table 3.7, all the fresh land plots, all the dry land plots, all the land plots) are significant for many indices. But regarding indi-

Table 3.6. Pearson correlation coefficients between *Erica arborea* FMC and AVHRR-derived indices on Les Maures (Year 1996, individual sites, MVC periods ending on Thursdays)

Plot type	Plot - series	Sample size	NOAA AVHRR-derived indices				
			NDVI	ARND	ST	NDVI/ST	WSI
Fresh sites	1	10	**0.68**	**0.70**	-0.24	0.44	-0.40
	2	10	**0.75**	**0.79**	-0.44	**0.66**	**-0.64**
	5	10	0.57	**0.80**	-0.59	**0.67**	**-0.70**
	7	10	0.49	**0.66**	-0.41	0.54	-0.53
	8	10	0.49	**0.62**	-0.28	0.40	-0.36
	10	10	0.52	0.61	-0.57	**0.63**	-0.58
	12	10	0.47	0.55	-0.48	**0.57**	-0.56
Intermediate	6	10	0.60	**0.77**	**-0.66**	0.71	**-0.63**
Dry sites	3	10	**0.77**	0.63	**-0.65**	**0.79**	**-0.72**
	9	10	**0.72**	**0.86**	**-0.72**	0.80	-0.44
	11	10	**0.85**	0.54	**-0.66**	0.78	**-0.69**
	13	10	**0.79**	**0.83**	**-0.78**	**0.88**	**-0.81**
	15	10	**0.85**	**0.84**	**-0.76**	**0.87**	**-0.79**

See typographic convention at the end of § 3.5.2.

vidual land plots (Table 3.6), the coefficient is now significant for different indices, particularly for the dry land plots and for a number of fresh ones.

Regarding index performance, the performance obtained with daily data is reproduced with MVC, but with more numerous significant coefficients. Globally for individual land plots, ARND is still the best performer, followed by NDVI/ST and NDVI. Then come WSI and ST. For groups of land plots, multi-temporal NDVI indices (GRNrel and SM), ARND and NDVI/ST are the best performers, but none is clearly above the others.

Table 3.7. Pearson correlation coefficients between *Erica arborea* FMC and AVHRR-derived indices on Les Maures (Year 1996, regrouped land plots, MVC periods ending on Thursdays)

Plot type	Sample size	NOAA AVHRR-derived indices						
		NDVI	GRNrel	SM	ARND	ST	NDVI /ST	WSI
Fresh plots	70	0.48	<u>0.53</u>	0.52	0.35	-0.35	0.46	-0.41
Dry plots	50	0.69	0.73	0.72	<u>0.78</u>	-0.71	0.77	-0.68
All plots	130*	0.40	<u>0.43</u>	<u>0.43</u>	0.42	-0.35	0.42	-0.38

* including land plot 6, with intermediate moisture
See typographic convention at the end of § 3.5.2.

3.5.6
Results on ONF land plots

Table 3.8 presents the correlation between NOAA-derived indices from MVC data and the foliage moisture of the most represented species in the ONF network. Results from 1997, the second year of operation of this network, are presented because the data analysis of the first year of operation (1996) resulted in 1997 in a greater homogeneity in the FMC data collection. This is due to the shifting of some of the 30 plots and a stricter application of the protocol. This was not an easy task for a network covering a total area of 84,850 km^2, and in an operation carried out by 15 different teams, one in each the 15 Mediterranean *départements* of the region.

The results are positive and coherent with all the preceding ones. All correlation coefficients, except 2 out of 48, are of the expected sign. In the same way, every species has at least one index in which its correlation coefficient is significant at a 95% level of significance. Two species, *Cistus monspelliensis* and *Erica arborea,* present correlation coefficients all of which are significant.

Regarding the comparative performance of the different indices, there are rather good performers in every family of index (vegetation indices, surface temperature or mixed), but none is above the others. ARND is still the best, but only with a short margin, and what is more, it is the one that presents the only two cases of incoherent correlation (sign opposite to expected). In contrast, ST, NDVI/ST and

NDVI are significant in 5 or 6 land plots out of 8, and the best performers on 2 land plots each.

Table 3.8. Pearson correlation coefficients between FMC of different species of the ONF network and AVHRR-derived indices (Year 1997, Maximum Value Composite data)

Species	Number of plots	Sample size	NDVI	GRN rel	SM	ARND	ST	NDVI /ST
Cistus monsp.	6	49	**0.35**	**0.30**	**0.32**	**0.49**	<u>**-0.52**</u>	**0.48**
Quercus ilex	4	35	<u>**0.48**</u>	0.32	**0.35**	-0.20	-0.39	**0.45**
Rosmarinus offi.	4	34	0.32	**0.40**	**0.40**	<u>**0.58**</u>	-0.55	**0.46**
Cistus albidus	4	33	0.15	0.02	0.05	-0.08	<u>**-0.71**</u>	**0.49**
Erica arborea	4	30	<u>**0.62**</u>	**0.36**	**0.45**	**0.50**	-0.45	**0.58**
Genista cinerea	3	24	**0.50**	0.18	0.27	**0.50**	**-0.50**	<u>**0.60**</u>
Quercus coccif.	3	23	**0.46**	0.37	0.40	<u>**0.59**</u>	-0.20	0.33
Juniperus oxyc.	2	18	0.30	0.03	0.09	<u>**0.63**</u>	-0.27	0.28

See typographic convention at the end of § 3.5.2.

3.5.7
Discussion

The results obtained are satisfactory, demonstrating that in a variety of land plots, a number of NOAA-AVHRR-derived indices are significantly correlated with the foliage moisture content of a variety of species. However, the results present a certain heterogeneity: for some plots or some species, no index was found significantly correlated, and it is difficult to single out the index that would prove to be the best overall performer, and as such could support an operational application.

Three main reasons can be found for this. A first one lies in some aspects linked to the change of scale. As explained earlier, the FMC protocol is a heavy one, particularly if you want to cover a certain number of plots during the 2 to 3-hour window corresponding to the maximum vegetation stress. The protocol adopted in this project, built on the experience of a team working on the subject since 1989 (Valette 1990, Moro et Valette 1996), has proved to be reliable. The measurements are made on 100x100 m plots, selected as being representative of a larger area. However, the soundness of this representativity has not been studied yet. In other words, the variability of the moisture of the vegetation within a NOAA pixel has not been tested. The foresters have developed a certain knowl-edge on the fertility of forest land plots, which can depend on many parameters like the nature of rock type, slope, aspect, etc. However, the spatial variations of these parameters cannot be known without intensive field campaigns. As a conse-quence the representativity of the 100x100 m plot in relation to the 1-km NOAA pixel cannot be ascertained. This aspect will need to be studied in a near future.

A second factor linked to the quality of AVHRR data has had consequences that amplify this problem of change of scale. It is linked to two characteristics of this sensor. The first concerns the variation in the size of the AVHRR pixel (from 1.1 km at nadir to around 2.5 by 6.5 km on the margins of the swath) and the second has to do with the residual imprecision in the accuracy of the location of the pixel. These characteristics make it necessary to compute indices, not on one pixel, but on a 3x3-pixel window, with the consequence that, due to the bigger area sensed, the problem of spatial heterogeneity is even more acute. In this respect, the use of the recently launched VEGETATION sensor should prove interesting, as it offers a 1-km resolution, constant across the swath, and an improved georeferencing, within 0.4 km, that would allow us to compute indices on one pixel, and not on a 3x3-pixel window.

A third important reason for obtaining bad results refers to the correlation obtained with MVC data. It is linked to the fact that the association of a measurement done on one day with an MVC computed on a 7-or 8-day period may end up with FMC and satellite data that do not match. This is particularly the case if the weather is variable during the week, resulting in different conditions of the vegetation. This was the case in France during the 2 years the project lasted, with unusually rainy periods during summer. Furthermore, AVHRR data taken on 7, 8 or 10 consecutive days are acquired with different view angles that cause bi-directional effects, i.e. noise, on data of the optical domain. Models that can correct a great part of these bi-directional effects have recently been developed, and additional research has to be conducted to test their use in improving of the quality of MVC, the need being the same with AVHRR as with VEGETATION data.

3.5.8
Conclusion

The results presented above, obtained within the framework of the EU-supported Megafires project, represent one of the first attempts to correlate daily acquired low resolution satellite data with a vegetation parameter critical in a forest fire context, the moisture content of living foliage, and all this on a significant European scale. This means that the validation of the satellite-derived indices was based on the measurements of foliage moisture content of different species carried out during 2 years on four networks of test-land plots situated in three Mediterranean countries.

From a methodological point of view, NOAA-AVHRR-derived indices have proved to be useful to monitor foliage moisture content as significant correlation have been obtained on the four test-land plots. Depending on the test-land plot, different indices obtained the best results: for example, ST and ST-linked indices (ST – AT, NDVI/ST) for Cabañeros, but ARND and multitemporal-NDVI ones for the Les Maures area. The interest of using daily or maximum value composite (MVC) has been more precisely studied on Les Maures. The analysis with daily images has, however, suffered from the scarce number of cloudless days, due to unusually rainy '96 and '97 summers. The analysis with MVC data, that, to a certain extent, provide clear data during cloudy periods, has shown good potential, showing significant correlations, especially when considering several land plots together. Some aspects, brought to light by the project, like the problem of "change of scale" between the size of field plots (100x100 m) and the size of

NOAA pixel (1x1 km) or even the NOAA 3x3-pixel window (3x3 km), should now be investigated. In this respect, the use of the recently launched VEGETA-TION sensor should prove interesting, with its constant 1-km resolution and its improved georeferencing.

From an operational point of view, this study has clearly shown that NOAA-AVHRR images contain information that could be used to monitor and provide some mapping of foliage moisture content. This would give information on a parameter that is considered important by fire fighting authorities, but that up to now could not be monitored conveniently. This parameter assessed from space, together with other parameters like meteorological ones, could be used for an improved estimation of short-term fire risk.

3.6.
Foliage moisture assessment using high resolution data

As mentioned earlier, high-resolution sensors have seldom been used in FMC estimation, since the temporal resolution and highcost of these images preclude their use in operational fire danger analysis. However, the use of high-resolution data is very convenient to calibrate methods generated for low-resolution images, since they simplify the procedure for locating control plots on the field because of the smaller pixel size. Additionally, high-resolution sensors offer information on the short-wave middle infrared spectrum (SWIR), which is the most sensitive to estimate water content. In the last NOAA satellite, a new band in the SWIR region has been added, but it has only been working since the summer of 1998 and during the morning acquisition, around 8.30 GMT, and therefore is still not very conven-ient for water stress assessment. However, Landsat TM bands 5 (1.55-1.75 μm) and 7 (2.08-2.35 μm) are placed just between the wavelengths of water absorption spectral regions (1.4, 1.9, and 2.4 μm), and should be well suited to determine water content. A number of studies have tested the usefulness of Landsat TM data in water content estimation, showing the potentials of SWIR data in this applica-tion (Cibula et al. 1992). More specifically related to fire danger estimation, other studies have tested the efficiency of thermal channels, following radiation budget models (Vidal and Devaux-Ros 1995), previously used in crop water content as-sessment (Moran et al. 1994). Results are very promising, since high correlation between these data and ignition delays were found for several Mediterranean shrubs.

As already mentioned, within the framework of the Megafires project, field measurements of FMC were performed in Central Spain. Although they were primarily addressed to assess multitemporal trends of NOAA-AVHRR data, three Landsat-TM images were also acquired and processed. They correspond to 23 April, 21 July and 23 September, 1997 (Plate 3.1). On these three dates, field samples were available. The first date had some haze contamination, and therefore the plots were extracted from neighbouring areas, with similar coverage and ter-rain characteristics to the areas sampled on the field.

The three Landsat images were geo-referenced and co-registered to ensure spa-tial coherence. RMS errors were in the range of 0.4 to 0.6 pixel. Calibration of raw digital values to radiance was performed using the coefficients included in the header files. The atmospheric correction was based on a recently proposed revision of the dark-object method, which estimates atmospheric transmissivity as a function of the cosine of the zenith angle (Chavez 1996). Conversion to reflectance included this variation on the simple method originally proposed for vegetation mapping (Pons and Solé-Sugrañes 1994). The thermal band (TM-6) was converted to radiance (L) using calibration coefficients included in the header. Surface temperatures were derived inverting Planck's equation (Wukelic et al. 1989). Emissivity and atmospheric effects were corrected with an energy balance method proposed by Vidal et al. (1996).

After obtaining the reflectances of the raw bands, several vegetation indices were computed to emphasise the discrimination of canopy water content. Indices included the Normalised Difference Infrared Index (NDII) (Hunt and Rock 1989), defined by:

$$\text{NDII} = \frac{\rho_{\text{NIR}} - \rho_{\text{SWIR}}}{\rho_{\text{NIR}} + \rho_{\text{SWIR}}} \ , \tag{2}$$

where ρ_{NIR} is the near infrared reflectance and ρ_{SWIR} is the short-wave infrared reflectance (computed either from bands 5 or band 7 in the case of Landsat-TM images). The NDVI was also computed, as well as the Water Deficit Index, and a ratio of NDVI and NDII to surface temperature.

All these indices were computed for five plots in which samples were collected. The average of a 3x3-pixel window was computed for each plot and image. Since only three Landsat-TM images were available for this analysis, quantitative correlation could only be computed for those species that were sampled in several plots in each period. This reduced the comparison to grassland (three plots per period = 9 observations), and one shrub species *C.ladanifer* (two plots per period = 6 observations), which is very common in Mediterranean areas. Additionally, an average of moisture content for all the shrub species (i.e. *Cistus ladanifer, Rosmarinus officinalis* and *Erica australis)* collected in the last two plots was computed.

Pearson r coefficients were calculated between FMC and all the indices presented previously, and with the raw bands of Landsat-TM (Table 3.9). From the analysis of the results, it is clear that indices based on the SWIR bands are better related to FMC than visible or NIR bands. Thermal data are also well related to grassland FMC, but poorly to FMC of shrub species, which are well adapted to the summer drought.

For grassland the correlation coefficient is significant with all variables except reflectance of bands 1 to 4. The trends of the correlation are logical, with negative signs for bands 5 and 7 (the higher the moisture content, the higher the absorptance, and the lower the reflectance), as well as band 6 (lower moisture content leads to higher temperatures and also implies higher surface temperatures), the difference of surface and air temperature, and the WDI, both of which are directly related to evapotranspiration. The other indices show positive correlation, because they refer to moisture content, while the NDVI is most probably measuring the changes in leaf area index and chlorophyll content as a result of grass curling. The cloudiness of the April image may contribute to the lower correlation found with indices based only on temperature data (ST-AT, WDI and B6), although they are also significant.

Table 3.9. Pearson r values computed between Landsat-TM variables and FMC of grassland and *Cistus ladanifer*.

Variable	Grassland		*Cistus ladanifer*		Shrub (Average)	
	R	Signifi-cance	R	Signifi-cance	R	Signifi-cance
B1	-0.034	0.930	0.121	0.820	0.223	0.672
B2	-0.139	0.722	-0.267	0.608	-0.223	0.671
B3	-0.585	0.098	-0.616	0.193	-0.593	0.214
B4	0.556	0.120	0.080	0.880	0.188	0.721
B5	**-0.915**	0.001	**-0.894**	0.016	**-0.869**	0.025
B6	**-0.796**	0.010	-0.350	0.497	-0.362	0.481
B7	**-0.884**	0.002	**-0.896**	0.016	**-0.877**	0.022
NDII$_5$	**0.965**	0.000	**0.886**	0.019	**0.905**	0.013
NDII$_7$	**0.941**	0.000	**0.900**	0.014	**0.921**	0.009
NDII$_5$/ST	**0.930**	0.000	**0.905**	0.013	**0.909**	0.012
NDVI	**0.963**	0.000	0.725	0.103	**0.748**	0.087
NDVI/ST	**0.975**	0.000	0.687	0.132	0.693	0.127
ST-AT	**-0.781**	0.013	-0.273	0.601	-0.288	0.580
WDI	**-0.758**	0.018	*0.117*	0.826	*0.091*	0.864

See typographic convention at the end of § 3.5.2.

For the shrub species measured, trends are similar to grassland although significant correlations are less numerous, both because of the fewer observations available (six cases) and the different behaviour of FMC for these species. The range of FMC values is much lower for shrub than for herbaceous species. *C. ladanifer* shows a 40% decrease in FMC (as a percentage of its dry weight) from the Spring to the Summer period, while grassland moisture decrease reaches 160% in the same periods. *C. ladanifer* is well adapted to the summer water shortage. Therefore, the FMC does not change abruptly along the season. Changes in leaf colour and leaf area index are much less obvious in shrubs than in grasslands, as well.

This may explain the low sensitivity of NIR reflectance to changes in FMC, although red reflectances offer a similar correlation with the water content to the one observed for grassland. Again, the best estimation of FMC comes from the SWIR data. Both bands 5 and 7 offer significant correlation with FMC, similar to the values measured for grassland. The composed indices based on SWIR also behave very consistently like the NDII$_5$, NDII$_7$, and MI$_{NDII}$, providing significant correlation with FMC.

As expected from the previous literature review, the NDVI and ratio of NDVI and ST do not provide as high correlation values as the SWIR indices. This is also applicable to grassland, since the values there were also significant, but for shrubs the indirect effects of changing FMC are less obvious than for grassland, and therefore, the NDVI is clearly less sensitive than the NDII to estimate them.

Indices based on surface and air temperature are not well related to FMC of shrubs. The differences in both temperatures are lower for shrub than for herbaceous species, especially in July and September. The WDI shows very a low correlation for both *C.ladanifer* and the average value of shrubs. This fact may be caused by the cloudiness of the April image, which severely obstructs the finding of the four corners of the trapezoid to compute the WDI (Fig. 3.2). However, it should also be pointed out that the method was initially developed for irrigated

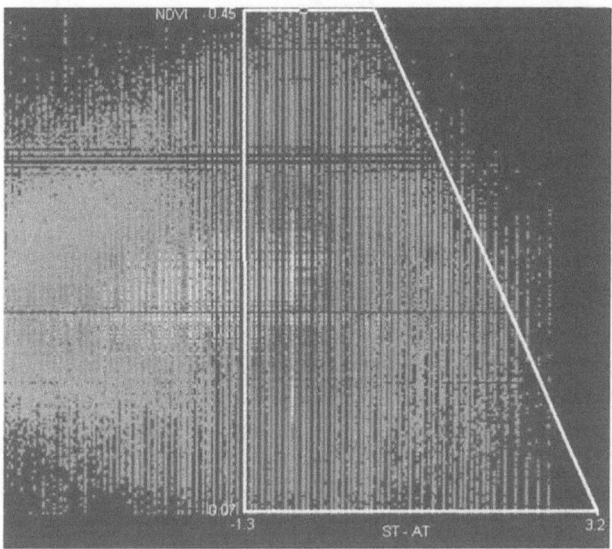

Fig. 3.2. Trapezoid of the April image derived to compute the WDI.

crops and therefore may require some adaptation for Mediterranean species. In any case, the small number of observations does not allow us to extract final conclusions on this lack of agreement between WDI and FMC.

For the most significant correlation, several fitting models were computed to explore the presence of non-linear trends, but no significant improvements were found with either logarithmic, exponential or growth models. Figure 3.3 provides some examples of fittings with linear and curvilinear approaches for grasslands and the *Cistus ladanifer*. The severe change in grassland FMC from the spring to the summer season is obvious from the analysis of the graph, because only two clouds of points are obvious in the images. Consequently, the assumption of non-linear tendencies does not have a solid basis for approval, and additionally qualitative evaluation provides better fitting for the linear trend.

Acknowledgements. This research has been funded by the Megafires project (ENV4-CT95-0256) under the Environment and Climate Program of the European Commission (DG-XII). Financial support was also obtained from the Spanish Ministry of Science through project number AGF96-2094-CE. Comments from Andrea Camia and Inmaculada Aguado, and linguistic help from Patrick Vaughan

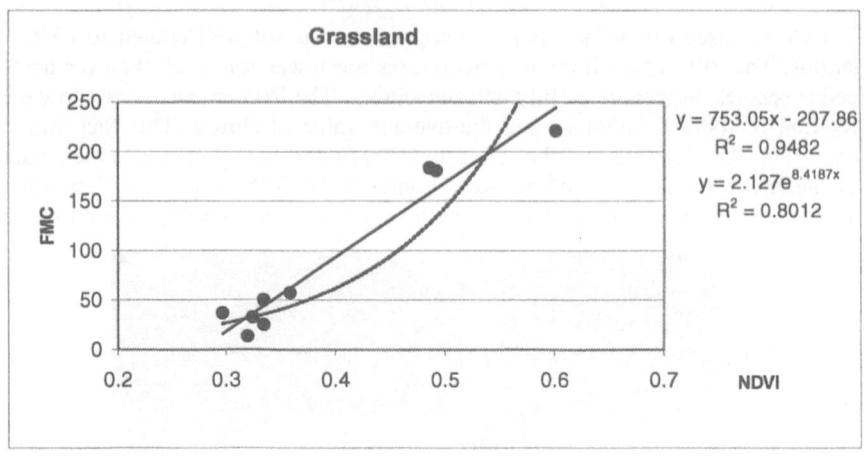

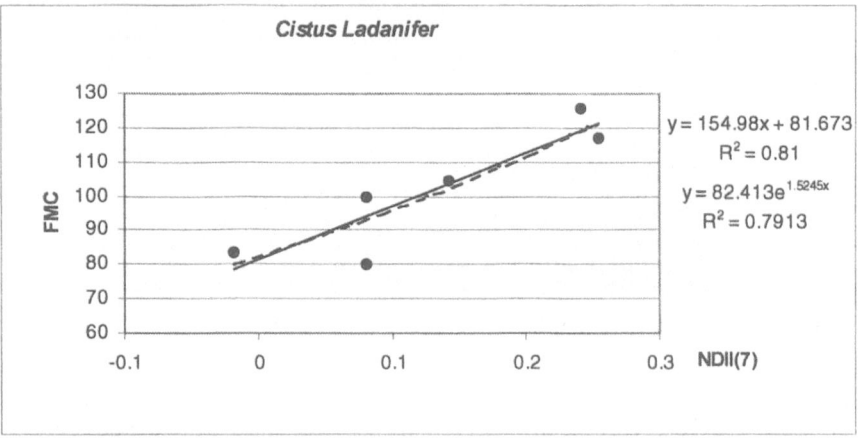

Fig. 3.3. Linear and exponential trends of FMC and NDVI and NDII$_7$ for grasslands (above) and *Cistus ladanifer* (below). Equations and R^2 are shown.

are also acknowledged. The director of Cabañeros National Park has also greatly contributed to the field work. We also wish to acknowledge the support of the *Délégation à la Protection de la Forêt Méditerranéenne* (MM. Grelu and Rivalin), INRA (JC Valette) and ONF (A Maillet), for granting access and supply us the results of the foliage moisture measurements campaigns carried out in France.

4 Meteorological fire danger indices and remote sensing

Andrea Camia, Giovanni Bovio
Department AGROSELVITER, University of Turin (Italy).

Inmaculada Aguado
Department of Geography, University of Alcalá (Spain).

and Nicolas Stach
CEMAGREF - ENGREF Remote Sensing Laboratory (France).

Abstract. Meteorological fire danger indices with a specific focus on large fire danger rating, and their potential integration with satellite data to improve spatial and temporal resolutions of the estimates are the themes of this chapter.

In this framework the main physical processes and components embodied in some fire danger indices that are mostly concerned with remote sensing will be outlined, and the main meteorological fire danger rating methods applied today in some of the fire prone areas of the world will be briefly recalled. Details will be given about a meteorological large fire danger rating model tentatively formulated for the European Mediterranean basin, recently developed within the framework of the Megafires project.

The use of satellite data to directly estimate some of the danger indices described will be explored, and some recent results obtained in this field both at regional and global (i.e. European Mediterranean basin) level illustrated.

4.1 Introduction

Meteorological fire danger indices are designed to rate the component of fire danger that changes with weather conditions, using a numerical or qualitative indicator derived by a combination of meteorological variables.

Weather and its changes over time affect both fire ignition and fire behaviour potentials, mostly through the fuel moisture content variation and the wind field properties, which are the basic variables involved in fire weather-related processes. Fuel dryness is strongly related to its flammability and combustibility, and consequently to fire occurrence and behaviour (Viegas et al. 1991), and wind plays a generally recognised key role during the flame front propagation (Rothermel 1983).

Weather conditions and fuel dryness have complex relationships that depend not only on the past and present status of meteorological variables, but also on the type and structure of vegetation and fuel layers and, furthermore, they can depend on other site-specific features, such as topography or some soil physical properties. Thus, the attempt to estimate dryness with weather variables has led in some cases to the development of indices that emphasise the modelling of specific phenomena, such as for example the moisture content of dead or of living material, the moisture content of specific fuel layers with a defined position in the vertical structure of the fuel complex or also the soil water deficit.

According to the way in which it is designed, a fire danger index can be more suited to rate fire ignition potential, for example focusing on those particles of the fuel complexes most prone to be initially ignited when a fire starts, or to address some of the other fire danger components, for example attempting to model the influence of meteorological variables on fire behaviour potential.

In the most complex fire weather indices, research has resulted in the definition of fire danger rating systems composed of various sub-models addressing specific components of the fire environment, thus providing estimates of both fire occurrence and fire behaviour potential. Actually, to completely assess wildfire potential, not only weather conditions but also fuel and topography data should be considered. This information has been included, with increasing spatial detail during last years, in the most advanced fire danger rating systems (Burgan et al. 1998, Stocks et al. 1989). In this direction, and due to the need of spatially and also temporally refining fire danger estimates, remote sensing techniques are expected to make important contributions in the near future.

In the first part of this chapter those physical processes embodied in the meteorological danger indices that are mostly concerned with remote sensing will be outlined. Subsequently, a brief recall of the main meteorological fire danger rating methods applied today in some of the fire-prone areas of the world will be given, to introduce the recent and initial definition of a meteorological large fire danger rating model for the European Mediterranean Basin, developed in the framework of the Megafires project. The last part of the chapter is dedicated to showing how and to what extent remote sensing can be used today to integrate these same meteorological indices, to improve the spatial and temporal resolution of the estimates.

4.2
Processes and components embodied in fire danger indices

The components that meteorological fire danger indices incorporate are mostly related to the moisture content estimations of dead and living fuels and their connected concept of drought. The role of wind, synoptic weather conditions and upper air layers are also often concerned in fire danger rating methods, but our main focus will concentrate on the above-mentioned issues, that have a closer relationship with remote sensing.

A detailed description of the physical processes involved in dead forest fuel moisture content variation can be found in Simard (1968) and Viney (1991). Precipitation effect, evaporation and vapour exchange between the fuels and the atmosphere (adsorption-desorption) are the processes mostly concerned in fire danger rating. With reference to vapour exchange, dead fuel moisture content changes continuously in time according to variations, mainly in air temperature and relative humidity, towards a theoretical moisture content of equilibrium with the atmospheric conditions, at a rate that is a function of the response time of the fuel particles. This concept derives from the assumption that the convergence towards the equilibrium moisture follows an exponential curve that can be divided into periods of equal relative amount of moisture exchange called timelag periods. The timelag of fuel particles is mainly controlled by their size or, in the case of ground fuels, by their depth within the soil layer. Hence fuel particles are often classified according to their size, as parameters indicating their approximate timelag value. The fuel size classes mostly used have a timelag of 1 h, also referred to as fine fuels, i.e. litter leaves and dead woody material with diameter less than 6.2 mm, 10 hours (diameter 6.2-25.4 mm), 100 hours (diameter 25.4-76.2 mm) and 1000 hours (diameter greater than 76.2 mm).

As mentioned, the response time is also influenced by the position of the fuel particles within the fuel complex, e.g. the organic matter of the upper soil layers has a shorter timelag than the duff in the deeper soil levels, mainly because of the local humidity regime. Furthermore, soil moisture influences the moisture content of the fuel particles above (Hatton et al. 1988).

The longest response time components of fire danger indices refer to the concept of drought, a condition of long-term moisture deficiency. The drought indices provide an insight into the seasonal trend of dryness, highlighting long-term moisture deficiency periods and disregarding the day-to-day variation. In some cases these indices include empirical based moisture content estimates at various soil depths or more or less complex soil water balance models. Drought is then related to the moisture content of longer response time fuel, organic matter in the deeper soil layers and heavier woody fuels, that ultimately are mostly closed to the fuel loading, that is to the amount of fuel available for combustion. In fact, while fine fuels are mainly responsible for carrying the fire front, and thus are more related to fire spread, heavier fuels affect mostly the total amount of energy released by a fire (Rothermel 1972). Meteorological danger indices that focus on fire spread potential rating add to the fuel moisture content estimation, in this case mostly referred to fine fuels, a stress on the wind effect on fire propagation.

Fuel moisture content of live plants is more difficult to estimate because it depends not only on weather trends, but also on plant physiology and on soil properties (Kozlowski et al. 1991). In other words it follows normally a seasonal trend, but it is also partially affected by long-term soil water deficit at the roots level. The moisture content of living foliage has an important role in crown fire behaviour. According to Van Wagner (1967), it follows a seasonal trend that changes with the age of leaves but has quite limited year-to-year variation due to weather both in conifers and in broadleaf trees.

More recent results (Haines et al. 1976, Olson 1980, Simard et al. 1989) show that moisture content of living vegetation can be partially modified by the soil water content at the roots level, thus seasonal trend and drought conditions should both be accounted for in the estimates.

Although beyond the scope of this chapter, some further weather features strictly related to extreme fire behaviour potential must be mentioned here for their importance. These are synoptic meteorological conditions with typical wind profiles (Brotak and Reifsnyder 1976, Brotak and Reifsnyder 1977, Brotak 1991) and atmospheric instability (Haines 1988), also included in some meteorological fire severity indices attempting to address the most extreme fire environment.

4.3
Meteorological fire danger indices

In Europe there is no uniform approach to fire danger rating and different methods have been developed in the different Mediterranean countries. Below, a synthetic description of the main indices applied will be presented in geographical arrangement.

In Portugal an index derived by the Nesterov Index (Nesterov 1949) is currently used. It is described in detail in INMG (1988) and is composed of three meteorological indicators that provide, respectively, an assessment of atmospheric conditions of the day (function of air temperature and dew point), a cumulative index that integrates the daily atmospheric conditions all along the fire season corrected by a factor function of the previous day's precipitation, and a final index that combines the previous two with the wind speed.

The method used in Spain is based on a model of moisture content of fine fuels, thus providing an estimate of the probability of ignition that can be eventually combined with the wind speed to allow a rough rating of fire spread potential (ICONA 1993). The moisture content parameterisation is a slightly modified version of the one embodied in the BEHAVE model (Rothermel et al. 1986).

In France, various methods have been developed for the Mediterranean part of the country. Most of them refer to a soil water balance model developed by Orieux (1979) to estimate the soil water reserve that is than combined with various meteorological algorithms of interest. The soil water content is modelled as a simple function of precipitation and evapotranspiration, and it is applied to account for the drought conditions throughout the fire season. The index proposed by Orieux (1979) is a direct combination of the estimated soil water reserve and the wind speed, while the index I87 introduced by Carrega (1990) adds to the previous variables the air temperature and relative humidity; furthermore it includes the phenological stage of living fuels and the estimate of surface water reserve of the upper soil layers. The method of Drouet-Sol (Sol 1990) is called Meteorological Numerical Risk and is based on the product of three coefficients (a soil water reserve factor, a wind factor and a false relative humidity factor) summed to a correction factor function of expected rate of spread of the flame front.

In Italy, two fire danger indices are used. The index most widely applied (Palmieri et al. 1992), especially in the Mediterranean part of the country, is the result of a slight calibration of the coefficients in the McArthur's Forest Fire Danger Meter (McArthur 1967). In the Alpine regions an index called IREPI (Bovio et al. 1984) has proved to be more suitable that assesses the departure of real from potential evapotranspiration as a measure of soil water deficit in the plant root zone.

Some studies were made in Minerve I and II European research projects to compare the fire danger rating capabilities of different meteorological danger indices, in the attempt to find a common model to be applied in the European

Mediterranean area, but the final results left some uncertainty about the best model to use (Bovio et al. 1994, Viegas et al. 1996).

In Australia, two meteorological danger indices are used for grassland (Grassland Fire Danger Meter – GFDM) and *Eucalyptus* forests (Forest Fire Danger Meter – FFDM). The meters, originally published as nomograms and subsequently expressed as equations (McArthur 1966, McArthur 1967, Noble et al. 1980), provide estimates of fire danger as directly and linearly related to the expected rate of fire spread in the two land cover types, using empirically derived relationships with meteorological data. Empirical equations to estimate fuel moisture content of grassland (Noble et al. 1980) and of the surface layer of *Eucalyptus* litter (Viney 1991) are also included.

As mentioned previously, in the most advanced fire danger rating systems the current trend is trying to incorporate with increasing spatial and temporal resolution fire behaviour prediction models to the solely meteorological fire potential estimates.

The most complex methods of fire danger rating are the Canadian Forest Fire Danger Rating System (CFFDRS) (Stocks et al. 1989) and the U.S. National Fire Danger Rating System (NFDRS) (Burgan 1988, Deeming et al. 1977, Deeming et al. 1974). Both systems are made up of separate components and incorporate also fire behaviour models, thus taking into account not only weather variables.

The weather-specific part of the CFFDRS is the Forest Fire Weather Index (FWI) System (Van Wagner 1987). The FWI values are computed from three basic and two intermediate codes, which refer to specific components of fire danger. The three basic codes account for the moisture content of different fuel layers and are called Fine Fuel Moisture Code (FFMC), Duff Moisture Code (DMC) and Drought Code (DC). These basic codes are combined to form two intermediate indices which are named Initial Spread Index (ISI), based on the FFMC and the wind speed, and Build Up Index (BUI) based on the DMC and the DC. The combination of ISI and BUI gives the final code, named FWI. All the codes of the Canadian system include a cumulative factor that accounts for weather conditions of the previous days. The first three codes rate the moisture content of fuels with different response times to changes in weather conditions (timelag), accounting respectively for short-term (FFMC), mid-term (DMC) and long-term (DC) dryness. The ISI is designed to represent the fire spread potential, the BUI the amount of fuel available for combustion. The product of the two gives the FWI that, properly scaled, provides an estimate of fire potential which is quite close to the Byram's fireline intensity concept (Byram 1959).

The U.S. NFDRS is a mathematical model divided into many sub-models, providing estimates of both fire ignition probability and fire behaviour potential (rate of spread and energy released) under given fuel, weather and topography conditions. Many of the weather-specific elements are intermixed with other components of the system and cannot be simply pulled out, with the exception of the semi-physical dead fuel moisture sub-models that are specifically designed for fuel particles sizes with timelag classes of respectively 1 hour, 10 hours, 100 hours and 1000 hours and that are based on equilibrium moisture content models (Bradshaw et al. 1983). In the U.S., other indices are used to rate wildfire potential integrating the NFDRS computations. The most interesting for our purpose is the Keetch-Byram Drought Index (KBDI) that accounts for the seasonal trend of dryness, representing the cumulative long-term moisture deficiency estimate of the organic material in the ground (Keetch and Byram 1968).

4.4
Large fire danger rating with meteorological indices in the European Mediterranean Basin

The fire danger indices previously described are not specifically tailored to assess extreme fire danger conditions. High intensity fires that may occur under severe danger circumstances pose a great challenge to fire managers in fire-prone areas. In most cases a minority of large fires interest the great majority of the total area burned in a territory, and this is verified also in the European Mediterranean basin, where "large fires" are considered those with a final area burned of more than 500 hectares.

For their strategic importance in wildfire management, both fire behaviour modelling and fire danger rating research efforts have recently been devoted to the study of such extreme events. With reference to the role played by meteorological variables in large fire occurrence, a review of papers can be found in Bovio and Camia (1997). A number of works give a description of meteorological danger conditions associated with major wildfire occurrence (Brotak 1980, Brotak and Reifsnyder 1976, Haines et al. 1976). In some cases the development of fire danger rating systems specifically focused on severe fire danger conditions has been addressed (Simard and Eenigenburg 1991, Simard et al. 1987). In Europe, an analysis of meteorological danger indices during 512 large summer fires has been carried out within the Megafires project and presented in Bovio and Camia (1998). In the same project a first global attempt has been made to develop a system for rating, from ground weather measurements, extreme fire conditions in the European Mediterranean basin. The system is empirical, i.e. it is derived from statistical analysis of historical fire and weather data, and is made up of a combined set of selected existing fire danger indices. As a number of different processes can have an important role in the extreme portion of the fire environment, different components of fire danger, in this case represented by a set of meteorological fire danger indices with specific features, have been embodied in a unique comprehensive fire danger index, to optimise the discriminating power of extreme danger conditions in the Mediterranean environment. In the next sub-paragraphs we will present a synthesis of the work that has been carried out and the main results obtained about meteorological large fire danger rating in the Megafires project.

4.4.1
Databases and danger indices

A global historical fire and meteorological databases were addressed for the whole European Mediterranean basin. With reference to the wildfire data, the fire seasons, from June to September in the Mediterranean environment, of five consecutive years (1991 to 1995) were analysed. In this time period 512 large fires were recorded in the five countries (Portugal, Spain, France, Italy and Greece) concerned. Meteorological variables required to compute fire danger indices were obtained from the database of the MARS Project (Monitoring Agriculture by Remote Sensing) implemented at the Joint Research Centre (JRC-Ispra Site). In this database, in the time period considered, daily ground weather data recorded at 12 a.m. from about 360 weather stations were interpolated on 1389 square grid cells of 50x50 km². The interpolated data were used, as the grid cell dimensions, the temporal resolution and the interpolation procedure applied at the JRC were considered suitable for this stage of the work. A detailed description of the meteorological database and the interpolation procedures can be found in Van der Drift and Van Diepen (1992) and in Van Der Voet et al. (1994), while a discussion about the specific features of both wildfire and meteorological databases used within Megafires project is in Bovio and Camia (1998).

In Fig. 4.1 the layout of the grid cells with the location of the large fires that occurred in the basin are showed.

Historical daily meteorological data extracted from the JRC-MARS database were daily maximum and minimum air temperature, vapour pressure, wind speed, rainfall, potential evaporation, solar radiation. With these weather data 28 fire danger indices were computed for each grid cell (daily values for the five fire seasons). The indices computed had different features, addressing specific components of fire danger. They included a number of dead fuel moisture content estimators for different response time fuels, from short-term (fine fuels) to drought estimators, the latter to catch the seasonal trend and therefore indirectly, to some extent, accounting for changes in live fuel moisture content, indices designed to rate fire spread potential, composite indices, i.e. embodying various danger components, and some models of equilibrium moisture content. A Visual Basic code was written to process meteorological data and compute the indices. The program is called MFDIP (Meteorological Fire Danger Indices Processor) and the compiled version, running under Windows 95, is included in the CD-ROM of the book. A

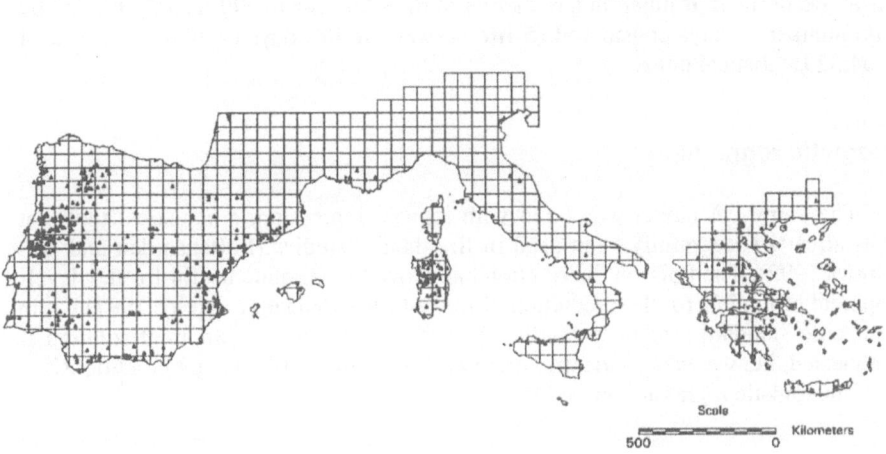

Fig. 4.1. Layout of the grid cells in the European Mediterranean basin and location of large fires in the fire seasons 1991 to 1995.

documentation file is also recorded with the descriptions of danger indices computed, the algorithms used in the program and related bibliographic references.

4.4.2
Climatic stratification

Due to the high heterogeneity of the fire environment in the territory, the area of investigation has been restricted and stratified using land cover and climatic data.

Those territories expected to be structurally only secondarily affected by large fires have been removed. To this end, the proportion of land cover classes of each grid cell was found with the CORINE Land Cover database and the grid cells mostly covered (more than 75%) by urban and agricultural lands were masked out. In addition the entire basin was stratified adopting the Köppen climatic classification system (Köppen 1936) to account for both environmental modifications due to climate and to different followup of the fire seasons. The climatic parameters required by the Köppen method were based on the same spatial resolution of 50x50 km², and were obtained processing a 22-year period (1975 to 1996) of daily meteorological data taken from the JRC-MARS database. The climatic classification obtained of the basin is presented in Plate 4.1.

Note that only the Köppen climatic types BSh, BSk, (semiarid types) Csa, Csb (temperate with dry summer types, i.e. Mediterranean) are truly representative of the investigated fire environment. This can also be verified by the fact that 502 out of 512 large summer fires recorded occurred in those areas (compare Fig. 4.1 and Plate 4.1) . The other climatic types have different conditions, some of them with a winter-spring fire season and a general lower level of fire danger. Hence only the climatic types Csa, Csb and BS, that aggregates the little represented BSh and BSk, have been retained for the development of the system.

The restriction of the European Mediterranean basin to the true area of interest as above defined, resulted in a reduction of the grid cells to 499, that multiplied by the number of days considered (5 fire seasons of 122 days each) give a total of 304,390 statistical units.

4.4.3
Logistic regression

The empirical model was build with logistic regression, a statistical tool that has already successfully been used in fire danger studies (Loftsgaarden and Andrews 1992). The Poisson distribution has proved to be quite useful for the development of models for the prediction of forest fires (Mandallaz and Ye 1997) but in our case standard logistic procedure is sufficient because a binomial solution is requested, i.e. we want to identify extreme / not extreme fire danger conditions.

The logistic regression model is

$$y = \frac{1}{1 + e^{-z}} \quad , \tag{1}$$

where y is the binomial dependent variable (values 0 or 1 in the data set), e is the base of natural logarithms and z a linear combination of n independent variables x_i

$$z = \beta_0 + \sum_{i=1}^{n} \beta_i \cdot x_i \quad , \quad (2)$$

where β_0 and β_i are the parameters to be estimated by maximum likelihood.

The explanatory variables we are looking for are given by the best combination of n fire danger indices that could explain the occurrence of extreme fire danger conditions. A more refined definition of extreme danger conditions with reference to the statistical units must be introduced at this point. Actually, large fires normally last several days, often burning most of the surface in one or a few highly severe days.

Unfortunately, wildfire data are not always recorded in such detail, and it was not possible to identify those specific days for all historical large fires in the data set. For this reason an approximation was introduced and an Extreme Danger Day (EDD) was defined for a statistical unit as a day with at least one large fire burning in the grid cell excluding, for multi-day burning fires, the day when the fire was extinguished, when it was reasonable to expect a lower level of fire danger. The dependent variable in the logistic regression equation is the binomial variable that identifies an EDD as defined previously, while the explanatory variables will be a selected set of danger indices that will provide the best fit.

Exploratory data analysis of indices led to identifying and excluding some indices with undesirable distribution properties. The following indices were discarded: IREPI showed too poor sensitivity to the change in weather conditions in the summer period, the ICONA and Orieux methods had distribution substantially equivalent on normal days and in EDD, McArthur's fuel moisture model (McArthur 1967) had too high a frequency of missing values due to the restriction requested in its calculation.

The number of possible combinations of indices to be tested has been further restricted to identifying pairs of indices mutually incompatible because highly correlated, in order to avoid redundancy and multicollinearity.

The logistic model was built using the backward elimination procedure of SPSS (Norusis 1993), testing several different initial sets of indices, until an optimum combination was obtained that, given the correct signs of the estimated parameters and the acceptable significance level of their coefficients, and verifying the low correlation between variables included, maximised the improvement in the Log Likelihood function.

Better results were obtained separating the three climatic zones, i.e. building different models in each zone and including indices of different types in each model (short, mid and long-term moisture content estimators, spread potential indices and composite indices), showing that when different components of fire danger are taken into account the estimates are improved.

The indices selected for each climatic zone and their coefficients in the logistic regression equations are shown in Table 4.1.

All models are globally significant at the 1% level and so are individual variable coefficients, with the exception of FM-Mark5 and IFDI in the Csb zone equation and RISNUM in the BS zone equation that are significant at the 5% level.

Note that the negative signs in coefficients are for indices that directly indicate moisture content values, while positive signs are for indices that increase with danger level. DC has both signs in the equation for the Csa zone, where it showed a quadratic relationship, but in the range of the index its overall contribution is always positive.

Table 4.1. Meteorological fire danger indices included in the three logistic models found for each Köppen climatic zone and their coefficients in the equations.

Meteorological Fire Danger Indices	Variables	Coefficients in the equations		
		Csa zone	Csb zone	BS zone
Fuel Moisture in Mark5 GFDM (Noble et al. 1980)	FM-Mark5	-0.014	-0.008	-
NFDRS 100 Hour Fuel Moisture (Bradshaw et al. 1983)	$(100\,h)^2$	-	-0.005	-
NFDRS 1000 Hour Fuel Moisture (Fosberg et al. 1981)	1000 h	-	-	-0.198
CFFDRS Drought Code (Van Wagner 1987)	DC	0.012	-	-
CFFDRS Drought Code	$(DC)^2$	-9.E-06	1.3E-06	-
Keetch-Byram Drought Index (Keetch and Byram 1968)	KBDI	0.003	0.003	-
CFFDRS Initial Spread Index (Van Wagner 1987)	ISI	0.044	-	-
Italian Fire Danger Index (Palmieri et al. 1992)	IFDI	-	0.083	-
Mark5 FFDM (Noble et al. 1980)	$(Mark5)^2$	-	-	0.027
Drouet-Sol Numerical Risk (Sol 1990)	RISNUM	0.068	0.092	0.048
	Constant	-11.096	-7.784	-3.744

The set of variables selected for the model of the BS zone is quite different from the previous two, but the limited number of large fires that occurred in this zone and the small improvement obtained in the Log Likelihood function, do not induce the expectation of a very good fit of this model to the data.

The indices finally selected should not be considered as being generally better than the others. Exploratory data analysis showed that many indices that are not included in the equations had good discriminating power. Those selected are "set" of danger indices that together are expected to improve the rating capabilities of meteorological severe fire conditions.

4.4.4
Assessment of the logistic model

To assess the model, observed and predicted EDD must be compared. To this end data have been grouped according to probability classes (i.e. levels of the dependent variable) and also according to climatic zone and month. Predicted and observed EDD in these groups were compared and the chi-square test of goodness of fit applied.

When grouping according to probability classes the chi-square test showed a good fit of the logistic models to the data in the three climatic zones, while grouping by climatic zone and month resulted in a poorer fit. A similar behaviour has already been pointed out by Mandallaz and Ye (1997) that showed how grouping data reflecting the calendar structure may result in an inflated chi-square value and thus in a poorer fit, due to the strong time correlation between EDD. Nevertheless, it must be also remarked that the high values of chi-squares were sometimes caused by a limited number of outliers. Plots of predicted versus observed number of EDD are showed in Fig. 4.2.

Concerning the day-by-day trend during the entire fire season, the logistic model is still interesting in the Csa zone. In Fig. 4.3 the daily observed and predicted number of EDD, i.e. the number of EDD computed over all the grid cells of the climatic zone Csa on a given day, are plotted for the entire period. It is clear that the peaks are not reached, but the general trend is mostly respected. In the Csb zone the day-by-day performance of the model changes according to the year, while in the BS zone, maybe because of the limited extension of the area and the lower number of large fires occurring than in the other two, the fitting on a daily basis has little correspondence.

Another criteria to assess the model prediction capability is to apply the performance score for fire danger indices introduced by Mandallaz and Ye (1997). In our case, for a given danger index to be tested, an index I is computed which is the sum, for all the N statistical units considered, of the product of the rank of the index value in the ith statistical unit, denoted by $rank(z_i)$, and a binomial variable I_i that takes value 1 if the ith statistical units was an EDD and 0 otherwise:

$$I = \sum_{i=1}^{N} rank(z_i)I_i \qquad . \qquad (3)$$

In case of an ideal deterministic prediction, the d highest values of the danger index would have been recorded in the d EDD so that we would have obtained the largest possible value of I as:

$$I_{max} = \frac{d(2N+1-d)}{2} \qquad , \qquad (4)$$

where d is the number of EDD occurred in the N statistical units considered.

In case of a random guess, i.e. danger index independent from EDD, the index would take the value:

$$I_{random} = \frac{d(N+1)}{2} \qquad . \qquad (5)$$

The performance score is then computed as

$$Score = \frac{I - I_{random}}{I_{max} - I_{random}} \qquad , \qquad (6)$$

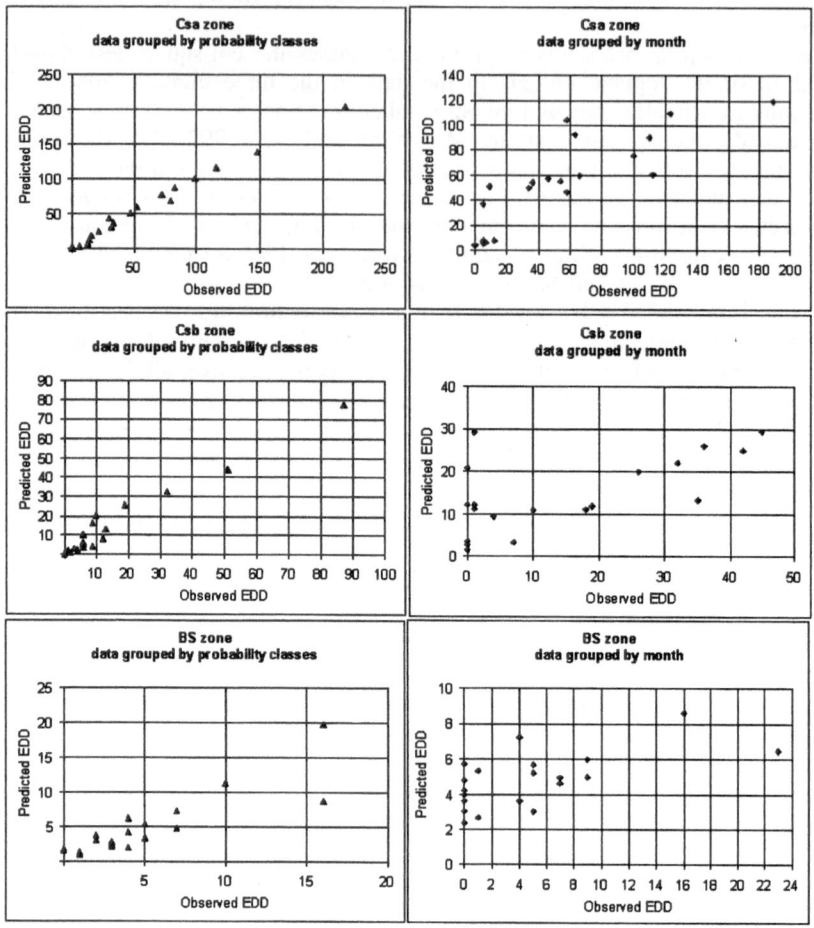

Fig. 4.2. Predicted versus observed number of Extreme Danger Days (EDD) in the three Köppen climatic zones.

that summarises the performance of the danger index with reference to a random prediction and an ideal, exact prediction. When *Score* is less than 0, the prediction is worse than pure chance (this could happen when the probability of the event decreases with increasing value of the danger index), 0 corresponds to the random guess score and 1 to an exact prediction.

In Table 4.2 the performance scores computed for a selection of danger indices and for the logistic models in the three climatic zones are shown. It can be seen that the logistic model had the highest score in all climatic zones, even though in the Csb zone some long-term moisture content indices (BUI and DMC) and the FWI also showed good performance, while in the BS zone the improvement obtained was lower, as could have been expected from the results illustrated previously.

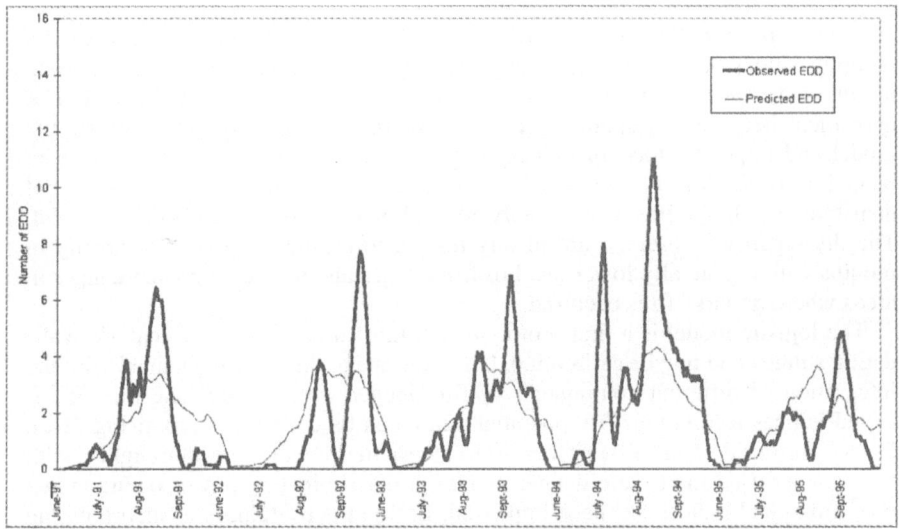

Fig. 4.3. Daily observed and predicted number of Extreme Danger Days (EDD) in the Csa zone during the fire seasons 1991 to 1995 (curves smoothed with 7-day running mean).

Table 4.2. Performance scores of a selection of meteorological fire danger indices and the logistic model in the three Köppen climatic zones.

Meteorological Fire Danger Indices	Variables	Performance Scores		
		Csa zone	Csb zone	BS zone
Logistic Models	LogModel	0.524	0.627	0.435
CFFDRS Fine Fuel Moisture Code	FFMC	0.321	0.541	0.414
CFFDRS Duff Moisture Code	DMC	0.305	0.577	0.195
CFFDRS Drought Code	DC	0.332	0.478	-0.161
CFFDRS Build Up Index	BUI	0.332	0.587	0.118
CFFDRS Initial Spread Index	ISI	0.297	0.466	0.391
CFFDRS Fire Weather Index	FWI	0.372	0.569	0.380
BEHAVE fine fuel moisture model	BEHAVE	0.323	0.541	0.350
NFDRS 10 Hour FMC	10 h	0.256	0.462	0.408
NFDRS 100 Hour FMC	100 h	0.260	0.475	0.421
NFDRS 1000 Hour FMC	1000 h	0.194	0.409	0.365
Keetch-Byram Drought Index	KBDI	0.312	0.290	-0.110
Mark3 (GFDM)	Mark3	0.363	0.321	-0.055
Mark5 Forest Fire Danger Meter	Mark5	0.277	0.488	0.384
Fuel Moisture in Mark5 GFDM	FM-Mark5	0.419	0.485	-0.016
Portuguese index	INMG	0.270	0.442	0.212
Carrega I87 Index	I87	0.299	0.490	0.234
Drouet-Sol Numerical Risk	RISNUM	0.320	0.516	0.365
Italian Fire Danger Index	IFDI	0.337	0.561	0.364

FMC, Fuel Moisture Content; *GFDM,* Grassland Fire Danger Meter

The spatial fitting of the model to the data can be qualitatively assessed by looking at the monthly maps of EDD locations in the month, plotted over the monthly values of the model output in each grid cell. These maps, one for each month of the time period considered, are recorded in the annexed CD-ROM. The agreement between the spatial distribution of the fire severity estimated by the model and large fire locations changes from year to year, but an overall correct behaviour of the logistic model can be recognised. Quite often in fact, the spatial distribution of large fires is reasonably grasped by the danger classes of the model. The discrepancies observed are mostly related to erratic large fires occurring in months with a generally lower level of fire danger and to overestimated danger in areas where no large fires occurred.

The logistic model is a first prototype that must undergo further analysis, validation studies and maybe calibration, but at this stage the results obtained with the integration of different components of fire danger with a composite mix of selected indices for extreme fire potential rating can be considered promising. Even though further data must be processed to assess the model more thoroughly, with the data set used the severe danger condition discriminating power of individual meteorological indices has been improved. Beside the usefulness of incorporating upper air data into the system, future refinements should include improvements in the estimates of the spatial distribution of variables. This is the direction of the work presented in next paragraphs.

4.5
Satellite data and meteorological danger indices

As illustrated in the previous paragraphs, traditional methods of fire danger rating rely on meteorological danger indices that take into account the critical variables of fire ignition and behaviour, and model their functional relationships on relevant weather variables. Besides the uncertainty inherent in the estimates of the models, the application of such indices presents some operational problems, since meteorological data for forested areas often are not available, and spatial interpolation techniques are not always suitable in areas of complex terrain.

In this context satellite data might represent a sound alternative, allowing some of the mentioned limitations to be overcome. In fact, satellite images are acquired through a systematic spatial sampling, the interval of each measurement being equivalent to the spatial resolution of the sensor. In addition, as the images are direct measurements at regular time intervals of land cover properties, some relevant parameters can be potentially obtained without the need of modelling their relationship with weather variables. Of course, not all the components embodied in meteorological fire danger indices are obtainable from satellite imagery, but especially those factors related to vegetation status, such as moisture content of both dead and living vegetation, can be potentially extracted and eventually integrated in fire danger rating systems with increased spatial detail.

The many studies carried out on the estimation of fuel moisture content from satellite data have been presented in Chapter 3. Such estimations in some cases could be used as components of those meteorological fire danger indices that incorporate this kind of information.

On the other hand, satellite data could be used for the direct estimation of the meteorological danger index (Domínguez et al. 1994, Walsh 1987). This was one of the objectives of the Megafires project and it will be the theme of this part of the chapter. It can be considered a further possibility in the tentative integration of specific components of fire danger indices into more spatially refined fire danger rating methods supported by the use of remote sensing techniques.

For operational use of satellite data in fire danger estimation, quantitative relationships between meteorological fire danger indices and satellite data need to be proved. Once consistent relationships are found, one should be able to use remotely sensed data for direct assessment in space of danger indices values or, alternatively, to compute some of their component codes.

Spectral data acquired by satellite-borne sensors are directly related to some critical parameters of vegetation in this respect. Strong absorbance of water in the middle infrared region (1.3 to 2.5 μm) makes this band most suitable for the estimation of water content of plants, as proven by several authors (Rock et al. 1986, Westman and Price 1988), and as such it is reasonable to hypothesise its potential usefulness in the case of meteorological fire danger indices estimations. Unfortunately, this spectral band is not included in most satellite systems currently available.

While future sensors, such as MODIS, may overcome this problem, the estimation of water content at present needs to be based on visible, near infrared and thermal infrared data. Indirect estimations of moisture content of living vegetation can be deduced from the red and near infrared bands, and therefore using spectral vegetation indices, which account for the contrast in these two bands. In fact, plant water stress implies a deterioration of leaf structure (Hale and Orcutt 1987) and, consequently, a clear decrease in near infrared reflectance and a light increase in red reflectance.

A commonly used spectral vegetation index is the Normalised Difference Vegetation Index (NDVI). Multi-temporal analysis of NDVI values has been shown to be valuable for rating forest fire danger (Burgan et al. 1998).

The main alternative to NDVI is based on the analysis of Surface Temperature (ST). In this case, water stress is estimated from plant evapotranspiration rates. When plants suffer water shortage, they reduce transpiration by stomatal closure and, as a consequence, the leaf is more heated (Kozlowski et al. 1991).

A combined analysis of NDVI and ST can also be approached, providing coherent estimations of fire danger, since dry vegetation offers lower NDVI values than wet vegetation and higher ST. The synthesis between the two has been based on simple ratios (Alonso et al. 1996), regression analysis (Nemani et al. 1993) and on the combined analysis of surface-air temperature and fractional vegetation cover (Desbois and Vidal 1996, Vidal and Devaux-Ros 1995).

As already mentioned, to better understand the relationships between such satellite derived indices and meteorological danger indices and explore their potential integration, studies have been carried out within the Megafires project and will be presented in the following.

4.5.1
Satellite data and the logistic model for large fire danger rating

The first experience presented was focused on understanding whether the logistic model illustrated previously, developed to refine extreme fire potential rating in the European Mediterranean basin, could be supported and/or integrated by satellite-derived information. To this end the relationship between the logistic model results and the satellite data was investigated at the basin level.

Obviously, as the logistic model was built for the Köppen climatic zones BS, Csa and Csb, which correspond to the most fire-prone areas of the basin and thus mostly concerned with large fires, also this analysis was limited to those climatic types. The logistic model was computed for a selection of weather stations of the JRC-MARS database and referred to circles of 10 km radius around them to avoid weather data extrapolation. A total of 56 weather stations was selected in the three climatic zones concerned, distributed in all the European Mediterranean basin (16 in the Iberian Peninsula, 19 in Italy, 16 in Greece and 5 in Southern France).

Regarding satellite indices, as they are expected to be more reliable on forest or shrubland areas, in the circular areas around the weather stations a mask of such cover types was applied using the CORINE Land Cover database. In addition, the isolated or too sparse pixels of this mask were eliminated through a shrink and expand process, to decrease the chance of incorrect matching due to mis-registration of images.

The AVHRR images used were acquired from NOAA-14. A set of cloud-free images was selected from early spring to fall 1996 for each region: 6 dates on the Iberian Peninsula, 10 on Southern France, 17 on Italy and Corsica, 16 on Greece. These images were georeferenced with the SGP4 orbital model. Through this standard registration procedure the position error can reach locally 3 or 4 pixels. Hence, in order to improve their geometric accuracy, the images on the Iberian Peninsula were co-registered with ground control points. On the other hand, to test the possibility of achieving comparable results with lower geometric accuracy at the basin level, the images of the other regions were not manually registered.

Daily NDVI and ST were computed around the selected weather stations and their correlation with the logistic model outputs analysed.

The expected signs of the correlation are positive with ST and negative with NDVI, and these signs were verified in the Iberian Peninsula images. However, a closer link was observed between ST and the logistic model (Pearson correlation $r = 0.67$), whereas the relation with NDVI appeared to be quite loose ($r = -0.27$).

As the logistic regression equations are actually different depending on the Köppen climatic zone, separate correlations were also considered for each one of them. The Pearson correlation coefficients of logistic models and ST range between 0.59 and 0.72 in the three climatic zones and are all significant at the 5% level, while for NDVI poor results were again obtained.

At the basin level, on account of the sparse relation between the logistic model and NDVI in the Iberian Peninsula, only the correlation with ST was assessed. Despite the lower geometrical accuracy of the images, reasonably good correlation was obtained, with r values in the three climatic zones ranging between 0.47 and 0.63, with the highest values in Csb zone.

In the time period from 19 to 23 July 1996 the strongest correlation were observed, and regression equations were developed with the logistic model output (LMO) expressed as a function of ST. For each Köppen climatic zone a separate equation was built of the form:

$$LMO = A \cdot e^{(B \cdot ST)} \qquad , \qquad (7)$$

with A and B being the estimated parameters and e the base of natural logarithms.

The r^2 values for the equations found were 0.64, 0.73 and 0.81 for Csa, BS and Csb zone, respectively . With such equations it was possible to produce a map of the logistic model output in the period considered, for the whole global area for which the logistic model was developed, using a temporal composition of the ST in NOAA-AVHRR images (Plate 4.2). This map is an example of what could be produced for other time periods once the relationships have been further consolidated.

4.5.2
Estimation of long-term fire danger indices from satellite data

While the logistic model integrates different meteorological components of fire danger, in the cases presented in this paragraph the potential use of satellite data will be explored for the spatial extrapolation of specific components of meteorological fire danger, namely the longest response time components, also referred to as drought indices. The indices considered are the Keetch-Byram Drought Index (KBDI; Keetch and Byram 1968) and the Canadian Drought Code (DC; Van Wagner 1987) illustrated in previous paragraphs. Designed to address the soil water content in the root zone and assess the seasonal trend of dryness, these codes are expected to be more related to the fuel moisture content of live plants than other components of meteorological fire danger. Hence, a closer association with satellite data is expected, since remotely detected radiation refers basically to the live vegetation canopy, being dead fuels normally hidden below, concentrated in the forest understory.

The analyses were performed at three different levels, in a region of Spain, in the Iberian Peninsula and in the whole European Mediterranean basin.

The regional study (Aguado et al. 1998) was performed in the Andalucía region (87,000 km²) located in Southern Spain, with data of 1995 and 1996. The AVHRR images used for this regional approach were acquired from NOAA-14 during the afternoon pass (around 14:30 local time). Spring images (from March to June) were also used to obtain information on the maximum vegetation vigour of each pixel. After solving reception and cloud-related problems, 91 images were made available for 1995, and 130 for 1996. All images were geometrically corrected and resampled to the UTM projection (zone 30, extended towards the western border) by using an orbital model. After navigating the images, an image-to-image registration was applied using control points, to insure consistency in the multitemporal analysis. A cloud mask was also applied to each image, based on multiband thresholding techniques.

As for the logistic regression study, the variables derived from NOAA-AVHRR data that have been tentatively related with fire danger indices are NDVI and ST. The NDVI is related to vegetation conditions, and its temporal variation follows basically plant physiology and therefore it is expected to show slow changes in time. Atmospheric scattering also has a strong effect on NDVI and to reduce it, daily images were synthesised using Maximum Value Composites (MVC).

The ST was calculated using a modified split-window algorithm (Coll et al. 1994) (revised by Coll and Caselles 1997). As in the case of the NDVI, after removing clouds from the image, the maximum value for each period was computed for ST, since we expected this value to be more related to high fire danger. This variable depends mostly on incoming solar radiation, cloudiness and wind, and to a smaller degree, on plant regulatory mechanisms. As a result, ST values will respond to atmospheric changes faster than NDVI values and consequently we hypothesised that long-term fire danger indices concerned in our study would have been better correlated to the latter.

Meteorological danger indices were computed from records provided by a number of automatic weather stations, selected because located near forested areas and providing consistent series of daily observations from February to September in the 2 years considered. The weather stations selected (27 for 1995 and 24 for 1996) were fairly well distributed throughout the study area, covering its main climatic districts.

Both temporal and spatial dimensions were addressed when analysing correlation of satellite variables with meteorological danger indices.

The temporal evolution of danger indices and satellite variables was correlated for each of the weather stations, to evaluate the performance of satellite data in estimating fire danger indices throughout the fire season in different regional climatic districts.

In addition, correlation within each 10-day period of satellite composites was analysed, in order to test whether the spatial diversity of meteorological danger condition could be predicted by satellite data in both spring and summer periods.

The temporal analysis showed good correlation of long-term danger indices with NDVI in both 1995 and 1996, but a poorer result for ST, preliminarily confirming our expectations. When considering the correlation in the space dimension, the same results were confirmed especially with reference to the DC-NDVI relationships.

Nevertheless, it should be noted that in our case association in the spatial dimension was generally lower than when considering temporal series. This may be due to the seasonal variability of fire danger indices which is higher than its spatial variability in the region considered. The same concept could be generalised by saying that the seasonal trend of vegetation in the Mediterranean environment implies a temporal contrast which is higher that its spatial contrast, and this is reflected also in the spatial and temporal distribution of long-term fire danger index values.

To summarise, a satisfactory correlation between the DC and the NDVI was found on a regional level. These relationships were proved in the temporal dimension and, to a smaller degree, in the spatial one. A regression analysis was also performed for the set of variables more strongly correlated, using DC as dependent variable and NDVI as explanatory independent variable.

Exponential functions of the form shown in equation (7) were found to provide a better fit. Separate functions were built for each weather station also separated by specific land cover types, splitting grassland, shrubland, and forest, with improved results especially for grassland, whose equations reached R^2 values of 0.63 in 1995 and 0.77 in 1996. Concerning shrubland and forest, the variability of the Mediterranean ecosystems in terms of species composition, and thus phenological cycles and responses to seasonal droughts, calls for further investigation.

These regression equations can be used for the estimation of DC values from NDVI with the spatial resolution of the images, without the need of interpolation.

Furthermore, DC maps can provide the spatial distribution of fire danger rated with the Canadian DC in areas where meteorological data are not available. Figure 4.4 shows an example of such tentative DC maps derived from NDVI data for the time period 11-20 July 1995 in the region of Andalucía. Agriculture, water, urban areas and the land cover type not covered by the regression equations have been masked out with the CORINE Land Cover database.

The possibility of generalising at a larger spatial scale the relation found between NDVI and DC has been further investigated at the other two levels of our study: first the Iberian Peninsula and then the whole European Mediterranean basin.

For this global approach both meteorological danger indices (DC and KBDI) were computed with records of the JRC-MARS database, while the satellite data were the same NOAA-AVHRR images used for the correlation analysis with the logistic model, with forested area masked out in circles of 10 km radius around selected weather stations (see previous paragraph).

In the Iberian Peninsula, temporal evolution of danger indices and satellite data were correlated for each one of 16 weather stations. In addition, as for the regional study, also the correlation in the spatial dimension was addressed.

In both cases the results of the regional study were confirmed, in most cases the correlation between NDVI and KBDI being quite poor, but the one between NDVI and DC was interesting, as will be detailed below.

In the temporal dimension 15 out of 16 sites showed the expected trend (i.e. negative correlation between NDVI and DC), with Pearson coefficients ranging

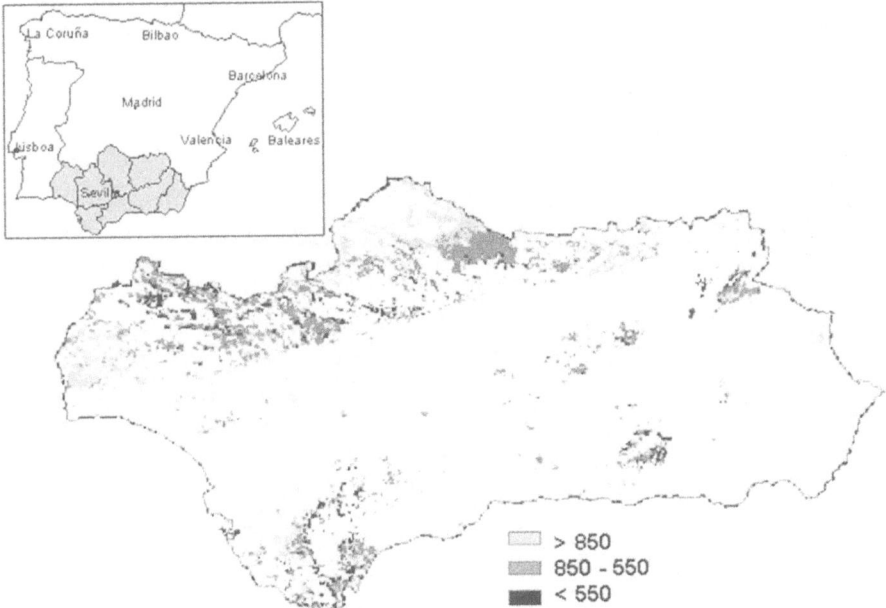

Fig. 4.4. DC values for grassland estimated from NDVI in the region of Andalucia (July 11-20, 1995).

from −0.12 to −0.91. However, due to the small sample size (6 dates), only 5 of these correlations were statistically significant at the 5% level.

In space, the correlation between NDVI and DC was significant for all the dates, also because of the larger sample size (16 observation in each date), with Pearson correlation ranging from −0.70 to −0.87.

As the results obtained for Andalucía and the Iberian Peninsula need to be tested and verified to be extended to the European Mediterranean Basin level, the analysis has been continued and repeated at different sites of the global area. The selection of 56 weather station of the JRC-MARS database and the mosaic of NOAA-AVHRR images mentioned in the previous paragraph was used. It should be noted that these images had lower geometric accuracy than the one used for the Iberian Peninsula, thus only some preliminary conclusion could be drawn from this global study.

The general trends obtained for Andalucía and the Iberian Peninsula could be observed at the Mediterranean Basin level. Regarding the temporal analysis, the

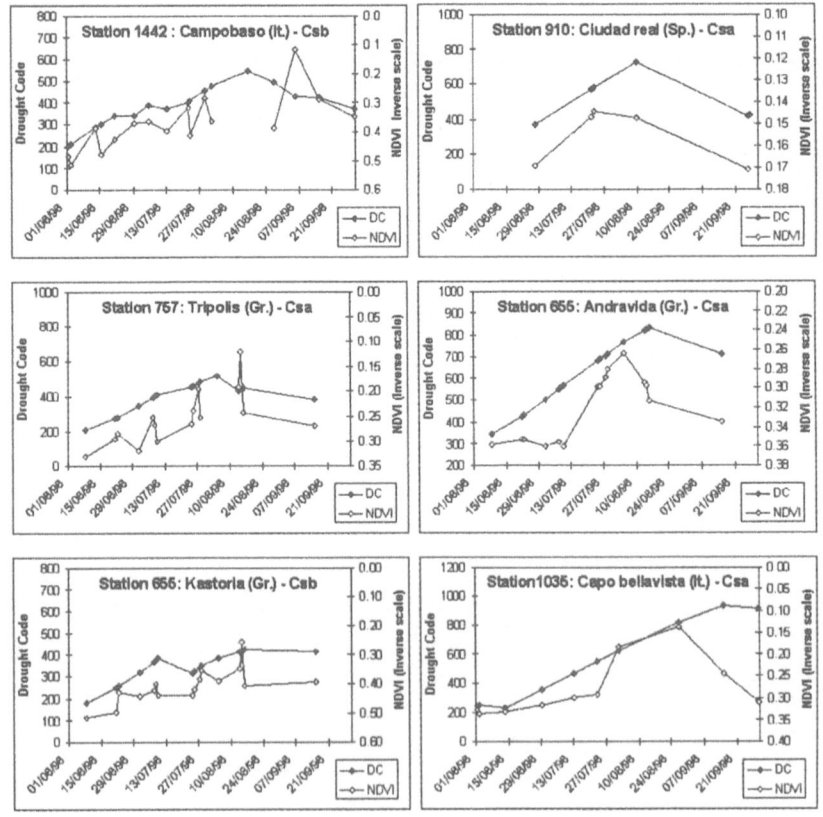

Fig. 4.5. Temporal profiles of DC and NDVI for a selection of weather stations during summer 1996.

expected signs of correlation between NDVI and DC were obtained for 44 out of 56 weather stations. Again, also due to the few dates available, the statistical significance at the 5% level could be established for only 9 of them. Figure 4.5 illustrates this inverse relation between NDVI and DC over a selection of weather stations.

Regarding the spatial dimension, the correlation between NDVI and DC calculated for all available dates showed the expected trend for 31 dates out of 35.

Restricting the analysis only on those dates with at least 10 weather stations available, 13 out of 20 dates showed a correlation between NVDI and DC statistically significant at the 5% level.

The strongest correlation period, from 19 to 23 July, with R=-0.74 and 92 records of available data, was then used to develop a regression equation to derive DC values as a function of NDVI. With the regression equation it was possible to produce a DC map from an NDVI temporal composition of the NOAA-AVHRR images of the same period (Plate 4.3)

The common background of some meteorological fire danger indices and remote sensing-derived indices was confirmed by observation at the regional level and partially at the European Mediterranean basin level, and this is particularly important for the long-term components of fire danger.

Furthermore, the DC, better than the KBDI, seems to be more related to living vegetation vigour and its seasonal trend in the Mediterranean environment, as it is monitored by remote sensing techniques. These consistent similarities, regardless of the local climatic differences, strongly support the use of satellite data to estimate temporal trends in some components of meteorological fire danger, and encourage their further investigation to profit from their potential integration into currently applied fire danger rating systems.

Acknowledgements. The Megafires Project was supported by the Environment and Climate Research Programme of the European Commission (DG-XII) (contract ENV4-CT95-0256 Climatology and Natural Hazards). The Joint Research-Centre – Space Application Institute, Agricultural Information Systems Unit and the MARS project are also acknowledged for having allowed the use of the data of the meteorological database. The European Environmental Agency is acknowledged for having provided the CORINE Land Cover database.

5 Integrated fire risk mapping

Emilio Chuvieco, F. Javier Salas
Department of Geography, University of Alcalá (Spain).

Luis Carvacho
Institute of Geography, Catholic University of Chile (Chile).

and Francisco Rodríguez-Silva
Environmental Agency of Andalucía (Spain).

Abstract. Fire risk may be considered from different temporal and spatial scales. This chapter reviews current methods to map fire long-term trends in fire risk by using several variables related to fire occurrence in a Geographical Information System (GIS). Several examples of both global and local-scale applications are presented. The former is focused on estimating the occurrence of large fires (above 500 hectares) from a set of human and physical-related factors. This analysis has been carried out at provincial level (NUT level 3) for Portugal, Spain, South of France, Italy and Greece. Logistic regression (LR) and Artificial Neural Network (ANN) techniques were used to estimate observed number of large fires from the geographical and statistical variables previously compiled. Local-scale studies are exemplified from the experience of the Environmental Agency of Andalucía, in the South of Spain, in which active use of GIS and remote sensing technologies is present in everyday fire management.

5.1
Temporal and spatial scales in fire risk mapping

Forest fire risk may be considered at different spatial and temporal resolutions: global and local; short-term and long-term. Both types of risk assessments are very important for fire management. Global scales can contribute to establish general guidelines for fire management at European level, while local scales are adapted to specific fire prevention resources of small regions. Short-term estimation of risk is required to take update decisions on fire pre-suppression and suppression activities, while long-term estimation addresses the general, more permanent, planning of fire fighting resources. The former should ideally provide daily estimations of fire risk and it is commonly based on weather data, although recently satellite information is also being considered (a review of such studies is available in Chap. 3 and in Desbois et al. 1997b).

This chapter is focused on analysing long-term trends of fire risk, those related to the more structural factors that affect fire ignition or fire propagation, such as topography, vegetation structure, human activities or weather patterns. These factors do not change daily but monthly or yearly, and can be considered stable at least during a whole fire season. This temporal scale is very useful to better understand spatial patterns of fire risk and to improve fire prevention management.

The most critical variables related to long-term fire hazard are climate (long trends of atmospheric variability), vegetation structure (height to surface ratio, compactness, flammability), terrain characteristics (slope, aspect, elevation) and human activities (land use, recreational practices), but they do not need to be updated frequently. Two to five year updates are accurate enough for fire management. Long-term fire risk maps are quite relevant for prevention and suppression purposes. They can help to design regional fire defence plans, which include fuel management and vigilance controls, such as fire-break design, dispatch planning, prescribed burning, look-out tower location, etc.

Since fire risk implies consideration of a wide number of cartographic variables, geographic information systems (GIS) are quite appropriate tools for this application. They provide spatial analysis capabilities and can easily integrate geographical variables. Fire risk GIS have been used in both spatial scales previously referred to, global and local, but local-high resolution systems have been more common (Yool et al. 1985, Chuvieco and Salas 1996, Chuvieco et al. 1997).

5.2
The use of GIS in fire risk assessment

A comprehensive consideration for fire risk implies taking into account a wide range of variables. A common terminology distinguishes between the concepts of risk associated to the beginning of a fire (fire ignition risk or flammability) and to the spreading of an active fire (fire behaviour risk or fire hazard: Salas and Chuvieco 1994). In each case, different variables and different risk weights should be considered. However, both approaches require being capable of integrating different spatial variables. GIS provide tools to create, transform and combine geo-referenced variables. Every analysis of geographical data with a GIS preserves the spatial dimension of variables being processed, because all transformations are performed cartographically. Therefore, GIS can spatially integrate several hazard variables related to fire risk, such as vegetation, topography, climatology and fire history, providing spatial tools for risk analysis (Aronoff 1989). Furthermore, these variables are spatially comprehensive (they cover the whole territory at the specific spatial resolution of the map information), which greatly helps forest fire management (Phillips and NicKey 1979). Consequently, GIS may portray the geographical location of those areas where risk factors are most severe and fire protection programs may be spatially and temporally oriented to the areas labelled as having high risk.

Most applications of GIS to fire risk mapping have been developed at the local level and therefore they cover a small area at high resolution (typically from 50 to 100 meter grid size: Bachman and Allgöwerm 1998, Castro and Chuvieco 1998, Chuvieco and Congalton 1989, Chuvieco and Salas 1996, Dagorne et al. 1994, Gum 1985, Helm et al. 1973, Langhart et al. 1998, Marchetti 1990, Salas and Chuvieco 1994, Salas et al. 1994, Salazar et al. 1990, Vliegher 1992, Woods and Gossette 1992,

Yool et al. 1985). However, there are also some experiences with global, low resolution, fire risk maps (Chuvieco et al. 1998, Miller et al. 1986, Werth et al. 1985).

For an in-depth analysis on the application of this technology to fire risk mapping, we will first consider the generation of the different risk and hazard variables, and then discuss the different schemes for their integration in a simple model of fire danger.

5.2.1
Description of geographical variables of fire risk

The applications of GIS to fire risk modelling have considered a wide range of hazard variables, depending on the specific characteristics of fire events in the different test sites. From the projects previously quoted, we may summarise the following list of variables:

- Topography (elevation, slope, aspect and illumination).
- Vegetation (fuel types, moisture content).
- Weather patterns (temperature, relative humidity, wind, and precipitation).
- Accessibility to roads and camping sites.
- Land property type.
- Distance to cities.
- Soils.
- Fire history.
- Water availability.

From all these variables, the most complex to generate is the vegetation-fuel map. Traditional vegetation maps are focused on the spatial delimitation of vegetation species following different taxons. From a fire management point of view, vegetation species are not primarily relevant for risk determination, since the same species may present completely different risk levels according to their morphology (height, density, compactness, ratio of dead to live elements), physiology (oil contents, moisture status) or landscape situation (vertical and horizontal continuity). Fire behaviour depends more strongly on these factors than on vegetation species (Albini 1985, Anderson 1982, Burgan and Rothermel 1984, Rothermel 1978).

In modelling fire behaviour, the great variety of vegetation characteristics related to the spreading of the fire has been summarised by the definition of different fuel types (Andrews 1986, Deeming et al. 1977). Fuel types may be defined as a classification of vegetation species according to their combustion properties (fuel loading, density, vertical continuity, compactness, area to volume ratio; Anderson 1982). Most fuel models have been developed considering that surface fires are the rule rather than the exception, and therefore they mostly consider the understorey component of forested areas. This is the case with the models developed for the Behave fire simulation program (Andrews 1986), which is extensively used in fire propagation modelling. This focus hinders the discrimination of fuel types from remotely sensed data, since remote sensors only obtain information from the upper canopy layer, with very little penetration capability. On the other hand, the average height and density of plants define some of these fuel types (for instance, Behave models 4, 5 and 6 are shrubs of different heights and densities), which are very difficult to discriminate from the spectral information gathered by remote sensors.

In spite of these difficulties, several papers have explored the use of satellite re-mote sensing to generate these fuel models through digital image processing. How-ever, visual analysis techniques have also been applied (Miller and Johnston 1985, Willis 1985). Most attempts are local-scale oriented and have worked with Landsat-MSS or TM images (Bradley et al. 1994, Burgan and Shasby 1984, Campbell et al. 1995, Castro and Chuvieco 1995, Cosentino and Estes 1981, Kourtz 1977, Rabii 1979, Root et al. 1985, Salas and Chuvieco 1995, Salazar 1982, Shasby et al. 1981, Vasconcelos et al. 1998). However, there are also interesting experiences of global fuel type mapping using low resolution sensors like the NOAA-AVHRR (Burgan et al. 1998, Miller and Johnston 1985, Werth et al. 1985, Zhu and Evans 1994). The accuracy measured by these studies ranges widely according to the different fuel types considered. Globally, an estimated accuracy of 65 to 80% has been obtained for Behave fuel types. Discrimination was particularly difficult between models 4 (high density and tall shrub, around 2 meters high), 6 (shrub between 0.6 and 1.2 meters) and 7 (similar shrub to model 6 but mixed with tree species) of the Behave program. They are spectrally very similar and usually require auxiliary information to achieve significant levels of accuracy (Salas and Chuvieco 1995). In the case of both Cana-dian-FBP and USA-NFDRS fuel models, which are more general than the ones used in the Behave system, accuracies are somewhat greater, ranging from 70 to 88% (Bradley et al. 1994, Burgan et al. 1998, Dixon et al. 1984, Root et al. 1986).

For digital classification of fuel types, topographic variables and texture bands have been commonly processed along with the original bands (Salas and Chuvieco 1995). More recently, neural network analysis and spectral unmixing has also been used for fuel type discrimination (Vasconcelos et al. 1998). Radar data might also provide complementary information for fuel mapping, particularly at local scale, since radar is very sensitive to temporal and spatial variation of the canopy (Churchill and Sieber 1991).

The human component of fire risk is critical in most Mediterranean countries, since human beings are the main agents of fire ignition, either by carelessness or arson. The spatial analysis of human risk is quite complex to model, since human activities related to fire are very diverse and difficult to be spatially represented. Therefore, to simplify the process two approaches may be considered: (i) deductive and (ii) inductive.

In the first alternative, human risk maps are created by overlaying several variables related to fire ignition. Some of these activities are spatially concrete, such as recrea-tion and burning of shrubland for pastures, which tend to be associated to particular areas of that cover. However, some others do not have a clear spatial pattern, such as arson, which is the main cause of fire ignition in some areas. Therefore, mapping all the human factors that may cause fire ignition is very complex.

The second approach for human risk modelling is more inductive, and tries to map human risk by relating location of fires to specific areas of land use or human activity. For instance, roads, camping sites, cities, urban-forest interface, or specific land use types may be related to fire occurrence (Alcázar et al. 1998, Chuvieco and Congalton 1989, Langhart et al. 1998, Martell et al. 1987, Vega-García et al. 1993, Vliegher et al. 1993). In the case of having access to fire perimeters of past events, they could be used to weight the different variables related to human activity by means of local adjustments, which may be based on multiple regressions (Castro and Chuvieco 1998, Chou 1992).

As far as climatological variables is concerned, the main problem to solve is the accurate spatial interpolation of single point observations (where the weather stations are located) to create grid layers. For instance, temperature or air humidity data are usually obtained from meteorological weather stations and therefore do not cover the whole territory. To obtain temperature or air humidity maps spatial interpolation techniques must be applied (Fujioka 1983, Hubbard 1994). The most frequently used techniques for the spatial interpolation of climate data are:

- Thiessen polygons.
- Weighted distance averaging.
- Kriging.
- Multiple regressions with auxiliary variables.

The first three procedures are based on distance, while the fourth assumes a strong relationship between the meteorological variable to be interpolated and the auxiliary variable (commonly elevation). The former are simpler to perform, but they do not account for the effect of topography (Burgan et al. 1998). Consequently, they are not very convenient, especially when the weather station network is very sparse and there are sharp contrasts in altitude. Extrapolation criteria might be used for the spatial distribution of air relative humidity, if some assumptions are held and a temperature map is previously generated (Chuvieco and Salas 1996). Wind data, on the contrary, is very difficult to interpolate, since wind flows are very difficult to model in complex terrain (Fosberg and Sestak 1986, McCutchan and Fox 1986, Ryan 1983).

5.2.2
Criteria to integrate forest fire danger variables

After creating the different risk and hazard layers to be included in the model, the most critical problem is to establish a coherent criterion to properly combine those variables. Since the goal is to obtain a single fire risk index, the component variables (vegetation, topography, weather, etc.) should first be classified in a numerical scale of risk and then combined into a single index. In some cases, the creation of risk levels from the original variables implies changing the nominal-categorical scale to an ordinal scale. For instance, different fuel types or slope ranges should be assigned a numeric value associated with a specific risk level. On the other hand, the integration of these layers in a single risk index requires that a weight be applied to each variable according to its importance on the fire occurrence (i.e. how much riskier are the fuel types than the slope?).

Both questions may be approached in a qualitative-subjective way or by using a quantitative-objective scheme. The former results from the experience of experts who assign risk levels and weights according to their own perception of fire risk in the area. The simplest way to develop this procedure is to create risk tables, where the combinations of two variables are assigned specific danger values (Brass et al. 1983, Gouma and Chronopoulou-Sereli 1998, Salas and Chuvieco 1994, Yool et al. 1985). Tables 5.1 and 5.2 include examples of such combinations.

The main problem of this approach lies on its subjectivity and local validity. Experts make the decisions, but even assuming their good knowledge of fire events in the study area, the method does not offer a clear rationale for extending the defined criteria to other areas. On the other hand, qualitative categories do not provide a clear image about gradients of risk presented in the field.

Table 5.1. Fire risk model proposed by Brass et al. 1983 (study area in Nevada)

Vegetation Slope	I	II	III	IV	V
0-9	Low	Low	Moderate	Moderate	Moderate
10-19	Low	Moderate	High	High	Very High
20-29	Moderate	High	High	Very High	Extreme
30-39	High	High	Very High	Extreme	Extreme
40 and more	High	Very High	Extreme	Extreme	Extreme

I, Agriculture, riparian, lush, grass; II, Sparse brush, sparse sage, sparse grass, hardwood, aspen; III, Jeffrey pine, pinyon/juniper < 30% of crown closure, cured grass, manzanita, medium density sage, mountain mahogany; IV, Jeffrey pine, pinyon/juniper 30-50% crown closure, dense sage; V, Jeffrey pine, pine > 50% of crown closure.

Table 5.2. Fire risk index proposed by Salas and Chuvieco (1994) (study area in Central Spain). This index is a combination of risk associated to fire ignition and fire spreading

Ignition Risk	Behaviour Risk			
	Very High	High	Moderate	Low
Very High	Very High	Very High	Moderate	Moderate
High	Very High	High	Moderate	Moderate
Moderate	High	High	Moderate	Low
Low	Moderate	Moderate	Low	Low

The quantitative approach to integrate fire-related variables can be achieved in different ways. First, it could be based on the selective weighting of danger variables to create single danger indices (Abhineet et al. 1996, Chuvieco and Congalton 1989, Lu et al. 1990, Salas and Chuvieco 1994, Vliegher 1992). Commonly these weights are based in the knowledge of the authors, in a way similar to the qualitative criterion, but this procedure does offer a gradient of risk levels, which can eventually be classified in different risk categories or used as they are produced. Some examples of such indices (Salas and Chuvieco 1994):

$$\text{Ignition Risk:} \quad IR = 4 * H + 3 * V + 2 * I - E, \tag{1}$$

where H represents human risk factor, V represents vegetation, I represents illumination factor, and E represents the elevation factor.

$$\text{Behaviour Risk:} \quad BR = 5 * V + 4 * S + 3 * A - E - FB, \tag{2}$$

where V represents fuel models factor, S slope factor, A aspect factor, E elevation factor and FB, the presence of fire-breaks.

These indices require establishing previously quantitative risk levels for each variable. For instance, fuel types should be ranked according to their ignition or behaviour risk. They are more objective than the qualitative criteria previously discussed, but they should be interpreted in a relative rather than an absolute way. In other words, they define higher and lower levels of fire risk, but they cannot be used to infer probabilities of fire ignition or rates of fire spread. Multicriteria evaluation techniques (MCE: Barredo 1996) may be a good alternative to reduce the subjectivity of this assigning process, since the opinion of experts may be quantitatively assessed. Moreover, each expert's opinion may be weighed according to his/her degree of knowledge in the field or the study area. MCE techniques have been used for fire risk mapping, weighing each risk variable after the expert's opinion in two different scenarios (Alcázar et al. 1998).

Another approach to obtain the weights of the different risk variables is using local regression analysis. Fire occurrence is the dependent variable, while fire risk variables are the independent ones. Coefficients of multiple regression become the weights of each risk variable for the synthetic risk map. Models proposed in the literature for obtaining these functions range from simple multiple linear regressions (Castro and Chuvieco 1998), to logistic regression models (Chou 1992, Vega-García et al. 1993) and neural networks (Chuvieco et al. 1998, Vega-García et al. 1993). Since these models are produced by statistical fitting procedures, the accuracy can be assessed quantitatively (that is the percentage of original variance explained by the model), and therefore a better understanding of the importance of each variable in fire occurrence can be obtained. However, they cannot be applied outside the period and study area where they were produced, and therefore these models cannot be extrapolated to other study areas.

Finally, the variables can be combined using outputs from standard danger indices or from fire behaviour models. Examples are available on the use of the US National Fire Danger Rating System (Agee and Pickford 1985, Lu et al. 1990), the Spanish ICONA's method (Chuvieco and Salas 1996), the Behave, Cardin and the Farsite programs (Caballero et al. 1994, van Wyngaarden and Dixon 1989, Vasconcelos and Guertin 1992, Woods and Gossette 1992). Using this approach, the risk maps produced may be more easily integrated in other phases of fire management. However, since fire behaviour programs are basically focused on fire propagation, fire ignition risk is not clearly included in GIS models based on those programs.

5.3
Analysis of long-term fire risk on a European level

5.3.1
Introduction

An example of the generation of integrated fire risk systems will be presented in this paragraph. The analysis is based on the results of the Megafires project, funded under the Environment and Climate Research Program of the European Union.

The main goal of this research was to identify factors of fire risk at European scale. Fire occurrence is focused on large events (fires above 500 hectares), which are the most critical from an environmental point of view.

The basis for this analysis has been the compilation of a database for the whole study area (Greece, Italy, South France, Portugal and Spain, with the exception of the Azores, Madeira and Canary Islands, respectively). The variables were extracted from national or global databases: Land cover CORINE, the Digital Chart of World, National Geophysical Data Center's Globe project, Defence Meteorological Satellite Program (DMSP) data. Census data have been the most difficult data to compile, since they present more consistency problems between the different countries. Census data have also marked the geographical unit of reference. Since most of the human variables were only available at provincial level (NUT-3), this division has been used for all the variables. In case of geographical layers covering the whole territory (such as land cover, elevation and DMSP data), the average value for each province was computed to assure consistency among variables. Within the study area, there are 164 NUT-3 provinces: 48 in Spain, 20 in Italy, 52 in Greece, 18 in Portugal and 26 in France. Sizes vary among the countries, being larger in Italy and smaller in Greece as an average.

5.3.2
Selection of risk variables

The selection of variables related to long-term trends of fire occurrence was performed from a previous literature review (Chuvieco et al. 1997), considering the limitations of data among the different countries. Three groups of variables were identified: geographical, dealing with terrain features (climate, land cover, roads, rails, etc.); demographic, related to population characteristics, and agricultural, referring to agricultural structure.

The analysis of census data in this project presented two limitations: the geographical unit of analysis, and the comparability between the countries involved. The first problem was addressed using the most common level of statistical spatial aggregation, that is the province. As for the second, some variables need to be discarded, either because they were not available in all the countries, or their meaning was not completely compatible and therefore comparisons would have been meaningless. This is the case of some variables that are, at least theoretically, related to human risk, such as property size, unemployment, and hunting practices. Since rural economies have strongly changed in most Mediterranean countries during the last 30 years, dynamic variables were also included in the analysis, by comparing present values with those measured in 1960. A whole set of 52 variables taken from demographic and agricultural census were extracted for the 164 provinces of the study area. The most critical are: Proportion of renters, Number and density of cattle, goats and sheeps, Proportion of agricultural area, Agricultural enterprises, Agricultural land (60-90), mechanisation, unemployment, population density, young index, aging index, active population, number of hotel beds, campings and proportion of second residences.

Regarding the geographical variables, a brief description of the spatial analysis undertaken follows:

- Elevation was obtained from two sources: National maps at 1:1,000,000 scale for Spain and Italy, and the GLOBE project (compiled by the U.S. Geological Survey at 1 km² resolution). From the elevation data (Fig. 5.1), mean slope and roughness was computed using algorithms provided by Idrisi G.I.S. (Eastman 1993).
- Land cover was extracted from the CORINE program, which is one of the layers generated for the European Environmental Agency (EEA 1996). The original coverage at 250x250 meters pixel size was degraded to 1 km² for the integration with other GIS variables. This degrading was performed by ap-

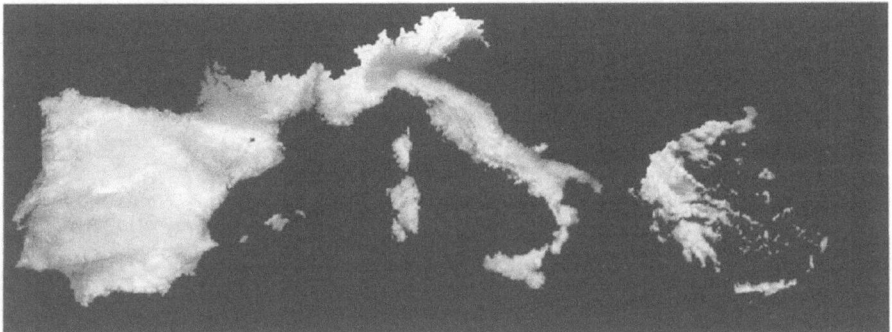

Fig. 5.1. Elevation of the Megafires study area.

plying a majority filter of 5x5 pixels to the original data and then a nearest-neighbour algorithm to the new 1-km² grid size. The original Lambert Conic projection was transformed to Albers Conical Equal-Area (ACEA), to be consistent with other data layers generated for the Megafires project (Vidal et al. 1997). The European legend of the CORINE land-cover program was simplified to six general fuel type categories: Grassland, Shrubland, Coniferous, Broadleaf, Agriculture, and Non-Vegetated (Plate 5.1). Finally, from each provincial unit, the percentage of each fuel type was retained for further processing.

- Density of roads and railways was computed from national maps and the Digital chart of the World. Both variables were originally coded in vector format. Consequently, the density calculation was also performed using the vector coverage of provinces using ARC/INFO (ESRI 1997). This software was also used to compute the centre point of each province (centroids), using standard tools provided by the software.

- Urban areas were assumed to be related to fire risk, since recreational uses of forests are some of the most important factors of fire ignition by accidents or carelessness. Since urban areas are very dynamic, satellite information was considered as a source to map urban land cover at global scale. Data from the Defence Meteorological Satellite Program (DMSP) provide a global view of city lights, since this satellite includes a very sensitive radiometer operating at night in the visible spectrum. The NGDC in Boulder, Colorado, process these data to generate a world database of city lights (Elvidge et al. 1997). City lights data from Eurasia were extracted from the Internet server at NGDC. The area of the Megafires project was delimited and converted to the ACEA projection (Fig. 5.2).

- Climatic regions were generated from the Joint Research Center climatic database archived at the MARS unit. This database provides daily values of various meteorological variables at a 50x50 km resolution. Average values for the last 30 years were used to classify each cell. Among the numerous climatic classification systems, the Köppen method was selected because it offers several advantages: it is simple, commonly accepted and based on bioclimatic criteria. The original Köppen system was reduced to fewer classes

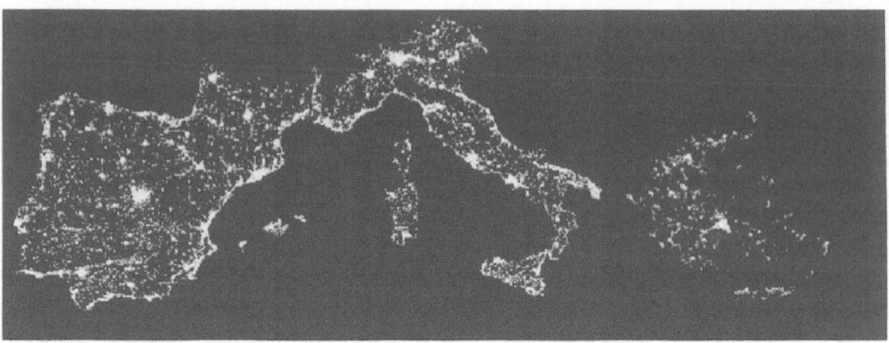

Fig. 5.2. DMSP city lights map of the study area (Source NGDC).

and those not presented in the Megafires study area were discarded. The final map includes the following climates: BSh+Csa, BSk, Cfa, Cfb+Cfc, Csb+Csc, Dfb+Dfc (Plate 4.1). The proportion of each provincial unit per climate type was extracted after cross-tabulation between the Köppen and the provincial GIS layers.

5.3.3
Techniques to estimate large fire occurrence

The number of large fires and burned area in each province of the EU Mediterranean countries were compiled from national statistics for the study period (1991 to 1995). This interval covers different trends of fire occurrence, 1991 and 1994 being the most disastrous, and 1992 and 1995 the most favourable. Distribution of large fires in the different provinces shows typical patterns of occurrence in the Mediterranean countries (Fig. 5.3). The most affected areas are the coastal regions in the Eastern part of Greece and Spain, the Central region of Portugal, Sardinia and Sicily in Italy, and Corsica in France.

Presence of large fires in each provincial unit was selected as the target variable to identify factors of fire occurrence from the risk variables previously selected. Two techniques were selected to perform this analysis: Logistic Regression and Artificial Neural Networks. As said before, both techniques have been previously used in the estimation of fire occurrence. Additional analysis was performed with the number of large fires per provincial unit. In this case, linear regression analysis was used instead of logistic regression, since this latter technique can only be applied to dicotonomous variables (fire/not fire). Neural networks were applied both to the presence or absence of fires and the number of fires per provincial unit.

5.3.3.1
Logistic Regression

Logistic Regression (LR) is a quite flexible tool, since it accepts the input of a data set composed of continuous and/or categorical variables as well as non-normally distributed ones. Several independent variables can be included in the

Fig. 5.3. Occurrence of large fires (above 500 hectares) in the study area (1991-1995).

model. Its main characteristic refers to the binary format of the dependent variable (0/1). In our case, the technique was used to predict occurrence/not occurrence of a large fire in each province, but it cannot be applied to estimate the number of large fires in each province. LR analysis is based on the following function:

$$f(z) = 1/(1+e^{-z}),\qquad(3)$$

where z is obtained from a linear combination of the independent variables estimated from a maximum likelihood fitting:

$$z = \alpha + \beta_1 x_1 + \beta_2 x_2 + ... + \beta\qquad(4)$$

where α is the constant and β_n, the weighing factor of the variable Xn. Z values can be interpreted as a function of the probability of occurrence. $F(z)$ converts z values in a continuos function that ranges from 0 to 1. Usually, LR values below 0.5 are assigned to non-occurrence of the independent variable (in our case, large fires), while values above 0.5 are considered as predictors of occurrence. However, this threshold can be modified according to the average value of the occurrence.

In order to avoid the effect of multiple correlation and the noise produced by the large number of variables in our study, a previous analysis to select the most significant variables was carried out. Some variables were normalised by using logarithmic transformations. Although it is not a requirement in LR, these transformations assure a more robust estimation as well as the possibility of comparing the results with other methods. Multiple correlations were computed after grouping the variables in three categories: (a) geographical variables (elevation, climate, land cover); (b) demographic (population growth, density and age structure, etc.), and (c) agricultural (density of land plots, rural enterprises, cattle, etc.). In the case of finding a pair of variables with Pearson r values above 0.5, only one was selected for subsequent analysis, to avoid colinearity effects and to simplify the process. After eliminating the most correlated variables, the following were selected:

- geographical: Bsh-Csa, Csb-Csc and Bsk Köppen climates, elevation, proportion of broadleaf, distance to roads, proportion of agricultural cover and density of lights;
- demographic: population density 90, difference in ageing index (90-60), difference in youth index 90-60, active population 90, difference in industrial population 90-60, difference in services population 90-60, unemployment 90, and difference in active population;
- agricultural: density of agricultural enterprises 90, difference in agricultural area 90-60, difference in agricultural companies 90-60, proportion of renters 90, differences in renters 90-60, density of cattle 90, density of sheep 90, density of goats 90, difference in size of agricultural enterprises 90-60.

These variables were input into three different LR, one for each of the above groups, in order to find out the most significant variables to explain fire occurrence. In all cases, the occurrence of large fires (0 not affected, 1 affected) was used as dependent variable. LR was performed with the Stepwise Forward Selection algorithm included in the SPSS statistical package (SPSS 1995).

Considering the geographical variables, the LR model predicted 79% of the large fire occurrence. The following variables were found significant: Bsh-Csa, Csb-Csc and Bsk (Köppen climates), elevation, % broadleaf, distance to roads and density of lights.

As for the demographic variables, 75.47% correct estimation was obtained and the following variables were selected: population density 90, difference in ageing index, active population 90, difference in industrial population, and difference in active population. All of them were significant at 95% confidence level.

Finally, the LR model with the agricultural variables provided a general agreement of 69.94%. The selected variables were density of agricultural enterprises 90, density of cattle 90 and differences in renters 90-60.

In order to build a global LR model, including the geographical, demographic and agricultural variables, a new correlation matrix was computed between them. Two variables, density of lights and density of agricultural enterprises, were discarded because of their close correlation with population density. Additionally, the three climatic variables were synthesised in just one, summing up the proportion of area covered by either Köppen's climate BSh, BSk or Cs. Consequently, 10 variables were finally selected for the LR model. As in the previous analysis the stepwise forward selection was used.

The final model was computed from 161 cases. Three provinces were discarded because of offering a high bias. The final model provided by LR was:

$$z = 0.8349 \ DENSI90 - 0.2168 \ POBACT90 + 0.6631 \ ALTIMED - 0.3524 \ BRLEAF + .0229 \ CLIM\_BC ,$$

where DENSI90 is the natural logarithm of population density; POBACT90 is active population 90, ALTIMED is mean elevation; BRLEAF is proportion of broadleaf trees and CLIM-BC is the provincial proportion with Köppen climates B and Cs. The signs of the coefficients are logical, since the fires are expected to be higher at greater population densities, higher elevations (rougher territory), more arid climates, with less active population and less area covered by broadleaf trees. The significance level is higher than 99% for active population, climate and population density, which should be considered as the variables most clearly related to large fire occurrence.

Table 5.3 offers an assessment on the performance of the model. A global accuracy of 78.26% was obtained in the estimation of large fires between 1991 and 1995. This prediction is quite high, considering the great diversity of the study area (geographical as well as national particularities). This performance is even better if only omission errors are considered, since only 11% of the provinces where fires occurred have not been estimated as such. Commission errors are higher (almost 40% of the provinces predicted as having a large fire were not affected), but these errors are less critical than omission errors from a fire management point of view.

Table 5.3. Observed versus predicted cases for the LR model (1991-1995)

	Predicted (cases)		Percent Correct
Observed (cases)	0	1	
0	38	25	60.32%
1	10	88	89.80%
Overall			78.26%

The parameters of the model are also quite significant. The log likelihoods are 223.19 and 146.22. These values can be used to derive a global index to estimate goodness of fit by using the following formula (Darlington 1980):

$$LRFC1 = \{exp[(LL_{model} - LL_0) / N] - 1\} / [exp(LL_0/N) - 1]. \qquad (5)$$

For this problem, the LRFC1 offers a value of 0.27, which is equivalent to a Pearson multiple r value of 0.52. The chi-squared value of the model is 76.98, while the threshold value at 95% confidence level is 11.1 (with 5 degrees of freedom). Therefore, the null hypothesis can be rejected with a high level of confidence.

The LR function derived from the final model makes it possible to compare the geographical distribution of expected versus observed occurrence of large fires (Fig. 5.4). Most provinces are predicted correctly. The main exceptions refer to the Piamonte and Lombardia regions in Italy and Navarra in Spain, which, in spite of possessing a low expected occurrence, have suffered several large fires in the period. In these cases, the effect of climate is quite obvious, since they present primarily an oceanic climate (and, therefore, low proportion of BS-Cs), which causes a low fire expectancy. It should be underlined that most fires affecting these regions occur mainly in winter or spring, and therefore they show a different pattern with respect to other Mediterranean regions.

The opposite effect, that is regions of predicted fires that were not actually affected, concerns some provinces in central Spain (Cordoba, Segovia, Soria, Palencia, Rioja), West of the Iberian Peninsula (Coruña, Braga), Southeast of Italy (Puglia), and West of Greece. In most cases, these are provinces with a high proportion of Mediterranean climate, medium to high population density and problems of unemployment. Although quite endangered, they were not affected by large fires in the period.

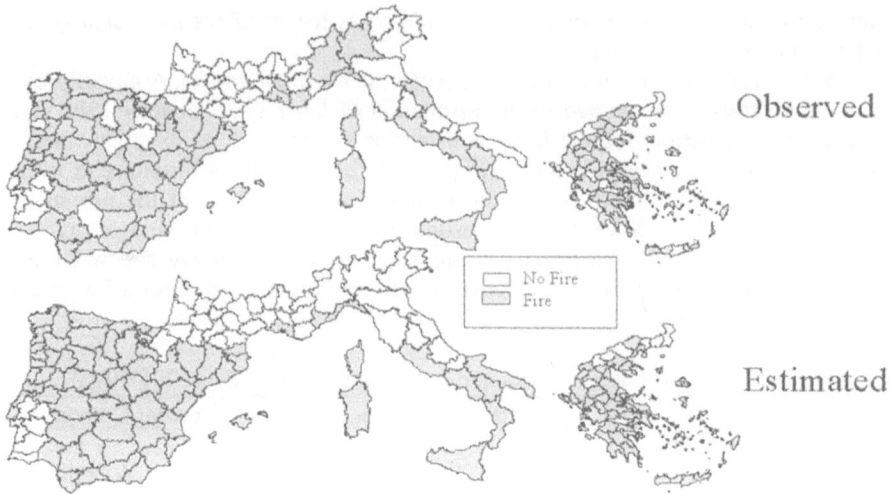

Fig. 5.4. Actual versus predicted occurrence of large fires by LR techniques (1991-1995).

To perform a final test on the LR model, a new equation was generated from a random sample of 60% of all the provinces. The other 40% of provinces were used as test cases. The LR model was created following a similar approach to the previous one. A threshold below 0.5 was also selected to discriminate between occurrence/not occurrence. In this case, as could be anticipated, the fitting is poorer that in the previous equation (Table 5.4). A global accuracy of 60% was achieved, which could be considered a good estimation, especially taking into account the rate of omission errors (37.5%). The same variables as the previous model were identified as significant to explain fire occurrence in these test provinces. Geographical distribution of the estimations is included in Plate 5.2.

Table 5.4. Observed versus predicted cases for the LR model. Test provinces

	Predicted (cases)		Percent Correct
Observed (cases)	0	1	
0	14	11	56.0%
1	15	25	62.5%
Overall			60.00%

5.3.3.2
Linear Regression

Linear regression analysis was applied to estimate the number of large fires in each provincial unit. Following a similar criteria to the logistic regression, a previous selection of significant variables in each thematic group (geographical, demographic and agricultural) was undertaking. Afterwards, exploratory analysis to the resultant variables was carried out to assure normal distribution of the independent variables. Some were log-transformed.

Multiple linear regression was computed for the final set of 30 variables. Criteria for selecting the independent variables were based on iterative stepwise forward method with thresholds of significance at 0.05. The final model retained was:

NF= 13.467 - 0.273 POBACT90 + 0.161 DIFPOBACT + 0.051 BSHCSA - 0.113 DENCAP90 + 0.048 DENOVI90 - 0.060 AGRICULT -0.073 DIFREN,

where NF is the number of large fires; POBACT90 is the percentage of active population in each province for 1990; DIFPOBACT is the difference in active population between 1960 and 1990; BSHCSA is the proportion of provincial area covered by Köppen climates Bsh and Csa; DENCAP90 is the density of goats for 1990; DENOVI90 is the density of sheep for 1990; AGRICULT is the proportion of agricultural cover in each province and DIFREN is the difference in agricultural renters between 1960 and 1990. The RMS error of this estimation is 5.4 fires. The highest errors correspond to Sardegna in Italy, Valencia, Castellón and León in Spain, Lesbos in Greece and Guarda in Portugal. All of them presented more than 10 fires of error in the estimation. These provinces, with the exception of Lesbos, had very high incidence of large fires in the study period.

5.3.3.3
Artificial Neural Networks

A neural network (or more properly Artificial Neural Network, ANN) is a network of many simple processors ("nodes"), each possibly having a small amount of local memory. Communication channels ("connections") which usually carry numeric (as opposed to symbolic) data, encoded by any of various means, connect the nodes. ANNs try to simulate the processes of learning carried out in the human brain, as long as we know how it works.

ANNs learn from examples and exhibit some capability for generalisation beyond the training data. This capacity has been extensively used in the past years to classify different types of data (Benediktsson et al. 1990, Bischof et al. 1992, Hewitson and Crane 1994, Openshaw 1994). The ANN does not have restrictions commonly found in traditional statistical techniques, such as normality or interval-scale information. Additionally, discrimination functions can be very complex, since the availability of different hidden layers makes it possible to define multiple connections between the input nodes, which do not need to be linear.

In order to achieve this learning process, ANNs are organised in layers (usually 1 to 5, albeit few of them need more than 2 or 3). Each layer has 1 or more "nodes" or "neurones", usually connected with all the nodes of the next layer. All the training process of ANNs deal with the calculation of the strength of this connection, for which purpose several algorithms have been developed, among which Quickprop and Backprop were tested in this analysis.

ANN structures were applied to the same estimations described previously. Both large fire occurrence/not occurrence and number of fires were predicted from the ANN. In order to make results comparable with the LR method, a similar set of variables was selected to train the ANN. In this case, instead of training three different networks (one for each group of variables), the same group of variables found significant after the preliminary analysis performed for statistical adjustment was selected. Consequently, a total of 26 variables was considered, among them forested area, population density 90, differences in active population, proportion of area under different Köppen climates, average elevation, proportion covered by broadleaf trees, proportion of agricultural area, density of lights, density of agricultural enterprises, agricultural area, density of cattle, goats and sheep.

In spite of the flexibility and fitting power of the ANN, one of the main drawbacks of this technique refers to the lack of general rules to decide the most convenient algorithm for training the network. A "trial and error" strategy is recommended by most authors, taking into account two performance criteria: reduction in the residual error and lack of "over-training". This last phenomenon can be detected when a validation set, running simultaneously with the training process, shows an increasing RMS, while the overall RMS of the training set decreases. This problem derives in an "overfitted net" and should be avoided. For the purpose of this project, the two aspects were considered. In all trials, 80% of cases were used to train the ANN and the remaining 20% to verify the accuracy.

The Quickprop algorithm provided the most accurate estimation of large fire occurrence. For the first case of estimating the fire/not fire occurrence for the whole period 1991-95, a three-hidden layer network was selected with 6, 1, 1 nodes (Fig. 5.5). In this case, the ANN predicts the 98.15% of the observed values, and all of them were correctly estimated. The fitting was obtained after 5,000 iterations, which provided an overall RMS error of 0.15.

Since the ANN are based in several hidden layers, where all the nodes are in-

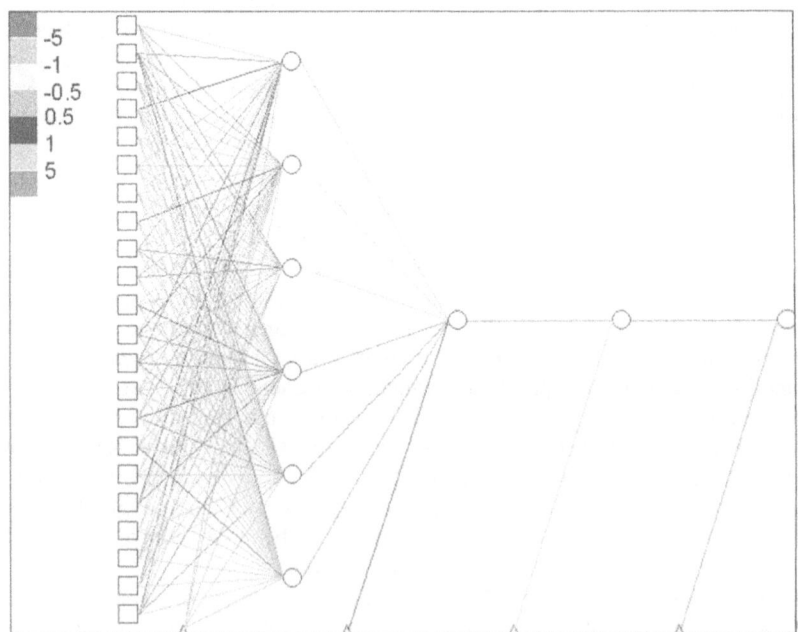

Fig. 5.5. Artifical Neural Network structure for the estimation of large fire occurrence.

terconnected, it is very complex to find out what are the most significant variables that affect fire occurrence. In this aspect, the ANN are like a "black box" with few analytical possibilities to measure the influence of the independent variables in the estimation. An indirect method to find out the most critical variables of the model was undertaken in the Megafires project. This method is based on replacing the original values of each input variable by random values, after the network is trained. We assumed that the increase in the RMS error produced by such a change should reflect the relative importance of that variable in the whole fitting. If the variable is significant, randomising it should give us a greater RMS with respect to a marginal one. Repeating this step with each variable showed the relative importance of all of them. Randomising Csb-Csc climates, average distance to the roads, active population in 1990, agricultural area, differences in population working in the services between 1960 and 1990 and population density provided the highest increment in RMS. Therefore, these should be the most critical variables in the estimation of fire-not fire.

Although this method may seem promising, some caveats should be stated. It is necessary to remember that the relations established by the network between the variables are not necessarily linear, and in fact, these relations may be different among variables. Consequently, changing the values of a single variable may affect the weight of others, which may be related with it. That variable can trigger the influence of a third variable perhaps strongly related with the previous one (with a function of unknown order) and over which we have no control in the experiment. Furthermore, we do not have the direction (sign) of the influence of each variable (positive or negative influence).

Following the same test performed for the LR model, a new ANN was trained for 60% of the provinces, in order to test the performance of the network for the remaining 40%. A different topology with respect to the one previously applied was used to train the ANN (1 hidden layer with 3 nodes). As in the case of the LR model, the prediction is considerably lower than for the prediction of the whole set of provinces (Table 5.5). However, the ANN still performs better than the LR model, offering an overall accuracy of 69%, and with much lower omission than commission errors (22 and 37.5%, respectively). The estimation is satisfactory for most of the provinces, with the exception of those affected primarily by winter fires (North of Spain and Italy) and some regions in central and northern Greece, in which actual fires were not predicted (Plate 5.2). Commission errors are located in Central Spain and Greece, with a high physical risk of being burned, but not affected by large fires within the study period.

Table 5.5. Observed versus predicted cases for the Artificial Neural Network model. Test provinces

	Predicted (cases)		Percent Correct
Observed (cases)	0	1	
0	17	12	52.62%
1	8	28	77.78%
Overall			69.23%

The ANN models are not limited, as LR techniques, to binary variables. Therefore, they can be used not only to predict fire/not fire, but also to estimate number of large fires in each province. This objective was pursued with a new training; in this case the output variable was the number of fires per province. After several trials, the lowest error was obtained with a back-propagation algorithm, using two hidden layers of 11 and 7 units, respectively.

Differences in estimated versus observed number of fires are quite low, with a maximum residual of 7 fires in the province of Barcelona. The residual mean squared (RMS) is 0.024, somewhat higher than the threshold. Scaling this value provides a residual error of 1.7 fires, which is much lower than the 5.4 fires or error measured from linear regression analysis of the same data. Using a standard value of 3 times the RMS, 98% of the cases are included in the estimations, since only 2 provinces (Barcelona and Trikala) present errors higher than 5 fires.

5.3.4
Conclusions

A draft estimation of global patterns of large fire risk across the European Mediterranean Basin was accomplished in this project. Two techniques, Logistic Regression (LR) and Artificial Neural Networks (ANN) were used to predict occurrence of large fires for the study period of 1991 to 1995. According to the Logistic Regression model the variables more clearly related to large fire occurrences are the proportion of BS-Cs climates, population density, elevation, unemployment and lack of broadleaf cover. ANN offers a robust estimation of fire occurrence, but it provides fewer insights into the most critical variables affecting the outbreak of large fires.

Estimated fire occurrence maps are a useful tool for managing global trends of fire risk, by pointing out those regions which offer a more consistent and stable risk of being affected by large events.

5.4
Examples of local-scale risk analysis

Local-scale fire risk studies are very useful for planning strategies for forest prevention through determination of spatial and temporal patterns of forest occurrence, as well as for the consideration of combustion potential of each fuel type and the study of man-made effects on natural forest (traditional agricultural practices, settlements, electric network, etc.). Furthermore, they can be the basis to assign priorities to the different tasks for fire prevention, like prescribed burning, planning of recreational activities, or building structures for (water reservoirs, road tracking, etc.). Finally, local-scale studies are used to support detection and extinction of active fires, by improving current network of watch-out towers, as well as the location of aerial and terrestrial fire fighting resources.

Among the different methods offered at local-scale studies, some of which have been reviewed previously, an integrated system currently being used by forest fire managers of the Andalucía region (South of Spain) is presented here in more detail. A brief review of its current application is also addressed.

5.4.1
Proposal for a local-risk index

The index is based on the analysis of the different factors related to fire ignition and fire spreading. The principal components of this index are shown in Fig. 5.6. The two main components are related to historical occurrence and potential danger.

Due to the spatial resolution of historical fire statistics, which are required to obtain the component related to fire occurrence, the spatial resolution of long-term fire risk is usually calculated at 10x10 km cells, and covers the whole region of Andalucía (87,000 km^2). However, the components related to potential danger can be obtained at much finer spatial resolution. Those components require digital elevation data, climatic information and fuel maps. The two former were obtained from the regional geographical information system (Sinamba) established by the Environmental Agency of the region. The Fuel Map was produced by direct assignation of forest categories of the land cover and vegetation map of Andalucía after intense field work. A total number of 46 forest categories were defined in the land cover map.

The historical occurrence or frequency index (Fi) can be defined in several ways (ICONA 1981). A simple way is by relating the number of fires in each cell to the number of years of the period considered:

$$Fi = \frac{1}{a} \sum ni ,\qquad(6)$$

where a is the number of years in the study period and ni the number of fires per year. This index is complemented with a causality index C, defined as:

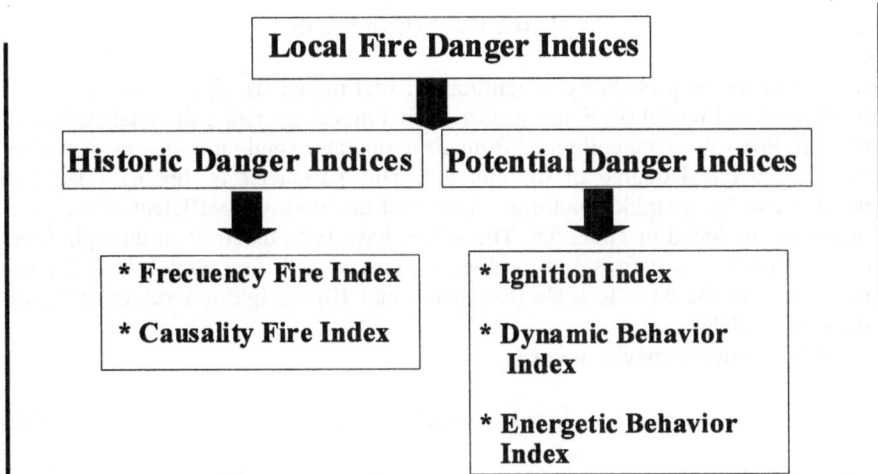

Fig. 5.6. Structure of the local-risk index proposed for the Andalucía region.

$$C = \frac{1}{a} \sum \frac{ci''.nmi}{ni}, \tag{7}$$

where a is again the number of years in the study period, ci'' the cause coefficient (weighted by cause type), nmi the number of fires by cause type, year and cell, and ni the number of fires per year and cell. The different causes are weighed according to the risk they historically generated: arson 10, negligence 5 and unknown 5.

The potential danger index, with independence of the historical analysis, provides an indicative about the risk associated to fuel type and condition, as well as topographic and atmospheric characteristics. This index, proposed by Rodríguez (1995 and 1998b), is based on three factors related to the ignition probabilities associated to each fuel model, the propagation properties and the energetic behaviour, once the combustion is consolidated. The information provided by these three components is well adapted to the three phases that characterise the organisation and consolidation of the combustion phase. Although the basis for calculation are 10x10 km squared cells, fuel data were compiled at 5x5 km resolution, by dividing each of the former cells in four squares. The three components of the potential danger index (PDI) are the ignition index (I_{ig}), the dynamic behaviour index (I_{cd}) and the energetic behaviour index (I_{ce}). All of them range from 0 to 10, and the final index is the summation of the three indices:

$$PDI = I_{ig} + I_{cd} + I_{ce}. \tag{8}$$

These components are defined as follows:
1. Ignition index (I_{ig}):

$$I_{ig} = \Sigma\, PM_i * (C_{ig})i * S_i\, /S, \tag{9}$$

where PM_i is the probability of ignition for fuel model (i); C_{ig}, the coefficient of ignition of fuel model (i); S_i the surface of fuel model (i), and S the total surface of the cell. PM values are computed from air temperature and humidity, corrected by slope, aspect and degree of illumination. This parameter is directly related to weather and topographic conditions. Values of the ignition coefficient of each fuel model are included in Table 5.6. The values have been determined through direct observation of forest fires and field assessment of each fuel model. Model 1 has been taken as the basis to scale (between 0 and 10) the ignition values of the remaining models.
2. Dynamical behavior index (I_{cd}):

$$I_{cd} = \Sigma P_i * S_i\, /S, \tag{10}$$

where P_i is an index computed from the rate of spread, and S_i and S are the same as in (9). Pi values are included in Table 5.7.
3. Energetic behaviour index (I_{ce}):

$$I_{ce} = \Sigma\, (Vp_i + Al_i + Ii + Cs_i) * Si/(4*S), \tag{11}$$

Table 5.6. Values of the ignition coefficient for diferent fuel models

Fuel type	C_{ig}	Fuel type	C_{ig}
Model 1	1	Model 8	0.5
Model 2	1	Model 9	0.4
Model 3	0.9	Model 10	0.2
Model 4	0.6	Model 11	0.2
Model 5	0.2	Model 12	0.1
Model 6	0.6	Model 13	0.1
Model 7	0.7		

Table 5.7. Values of propagation index for different ROS

Rate of spread (ROS) in m/min	(Pi) values
0-10	1
11-20	2
21-30	3
31-40	4
41-50	5
51-60	6
61-70	7
71-80	8
81-90	9
>90	10

Table 5.8. Energetic behaviour index values for different rate of spread conditions

ROS (m/min.)	L flame (m)	I Kcal/m/sec	CS Kcal/m^2	I_{ce} index
0-10	0-0.5	0-334	0-2090	1
11-20	0.51-1.0	335-752	2091-4180	2
21-30	1.10-1.5	753-1087	4181-6270	3
31-40	1.51-2.0	1088-1421	6271-8360	4
41-50	2.10-2.5	1422-1756	8361-10450	5
51-60	2.51-3.0	1757-2090	10451-12540	6
61-70	3.10-3.5	2091-2424	12541-14630	7
71-80	3.51-4.0	2425-2759	14631-16730	8
81-90	4.10-4.5	2760-3093	16721-18810	9
>90	>4.5	>3093	>18810	10

where Vp_i represents the index that depends on the rate of spread of the fire, Al_i is a length-flame index, I_i is related to fireline intensity, Cs_i is and index related to heat by unit area, and S_i and S are the same as in Eq.(9). Table 5.8 includes the main values of the I_{ce} index for different rates of spread (ROS).

5.4.2
Application at local level

The Potential Danger Index (PDI) is being utilised operationally for the fire management planning in the region of Andalucía. The value of the PDI can be computed to study the spatial or temporal distribution of risk. The former is used to better locate long-term fire prevention resources (look-out towers, fuel management), while the latter is used to plan specific actions in higher danger periods.

For testing purposes, results on the computation of this index have been restricted to the Natural Park of Los Alcornocales, located in the South of the Region. Although the basic management cell is 10x10 km for the limitation derived from fire statistics, the computation of the PDI can be performed at much finer resolution, since fuel, meteorological and topographic information is available at very detailed scales (1:50,000 to 1:100,000). For the area of los Alcornocales, the calculation of the PDI has been performed at 5x5 km cell size, although the basic information for fuels and topography is at 20x20 meter cell size. Table 5.9 summarises the input parameters to compute the danger values of the different fuel types in each of the four management cells (5x5 km), which cover the global cell of 10x10 km. The spatial diversity of the different indices, related to topographic or fuel characteristics, is obvious. Consequently, the spatial differentiation of danger values makes it possible to focus prevention measures on the more risky areas of the Natural Park, or in those areas that would suffer greater damage in the case of being affected by fire. In both cases, the spatial information is critical for the fire manager, in order to reduce the impact of fire on very valuable areas.

The meteorological data to compute the PDI for the Natural Park proceed from three weather stations located in the surroundings. Spatial interpolation techniques are used to obtain values of temperature, air humidity and wind speed for the centre points of each management cell. Weather-related danger is computed three times a day at 8, 14 and 20 hours. The final calculation of the PDI for the four management cells of Table 5.9 and the three periods previously referred to are included in Table 5.10.

It is important to emphasise that danger values change throughout the day, obviously not only within each cell, but also in the relation between the cells. For instance at 8.00 h cell B4b is the riskiest, while at 14.00 and 20.00 h, the most dangerous is B4c. This is critical for the fire manager, since fire prevention measures may be spatially and temporally optimised.

Figure 5.6 shows the fuel types of a 10x10 km cell of Los Alcornocales natural park, in which an important fire of 937 hectares occurred in 1997. This event may serve as a general assessment of the local-risk method proposed in this section, since PDI values were close to the high-risk level when the fire occurred (30.64 while the threshold for high risk is 32.87). These values were computed at noon (12 h). They were mainly related to fuel conditions, and not so much to meteorological risk, because wind speed was medium during the fire event. For this reason, fire managers were not completely aware of the risk before the fire started. Integral consideration of fire danger may improve current planning, by taking into account other factors more associated with local conditions (live and dead fuel moisture content, topography, and so on).

Table 5.9. Input parameters to compute the PDI of a sample of fire management cells covering the natural park of Los Alcornocales (Andalucia, Spain)

Sub-cell	Fuel Model	Number of cells	Surface (Hectares)	Si/S	Max Slope (%)	Mean Slope (%)
B4a	2	8450	338	0.1352	34	12.3
B4a	3	1069	42.76	0.017104	30	7.6
B4a	4	35	1.4	0.00056	35	23.7
B4a	5	120	4.8	0.00192	21	10,1
B4a	7	12407	496.28	0.198512	40	14,4
B4a	8	39985	1599.4	0.63976	42	16.6
B4a	999	434	17.36	0.006944	42	19.9
B4b	2	10013	400.52	0.160208	33	11.2
B4b	3	1563	62.52	0.025008	24	9.1
B4b	4	151	6.04	0.002416	20	6.3
B4b	5	112	4.48	0.001792	31	14.3
B4b	7	19573	782.92	0.313168	47	14.9
B4b	8	30736	1229.44	0.491776	40	16.2
B4b	999	352	14.08	0.005632	43	18.4
B4c	2	1829	73.16	0.029264	34	12
B4c	3	3505	140.2	0.05608	28	8.1
B4c	4	129	5.16	0.002064	25	12.5
B4c	5	733	29.32	0.011728	24	10.2
B4c	7	28192	1127.68	0.451072	42	13.2

Table 5.10. PDI values computed for four adjacent cells and three different hours in a single day

Hour	PDI, B4a	PDI, B4b	PDI, B4c	PDI, B4d
8.00	3.68	3.64	4.75	4.15
14,00	5.28	6.09	6.68	5.6
20.00	2.92	3.22	3.72	3.41

Obviously, this index can be computed for instantaneous evaluation of fire risk, as well as for general evaluation of long-term trends, which are critical for better management of fire prevention resources.

Acknowledgements. Different groups within the Megafires consortium participated in the generation of the database for the global study of large fire occurrence: Instituto Superior de Agronomía, in Portugal; University of Torino, in Italy;

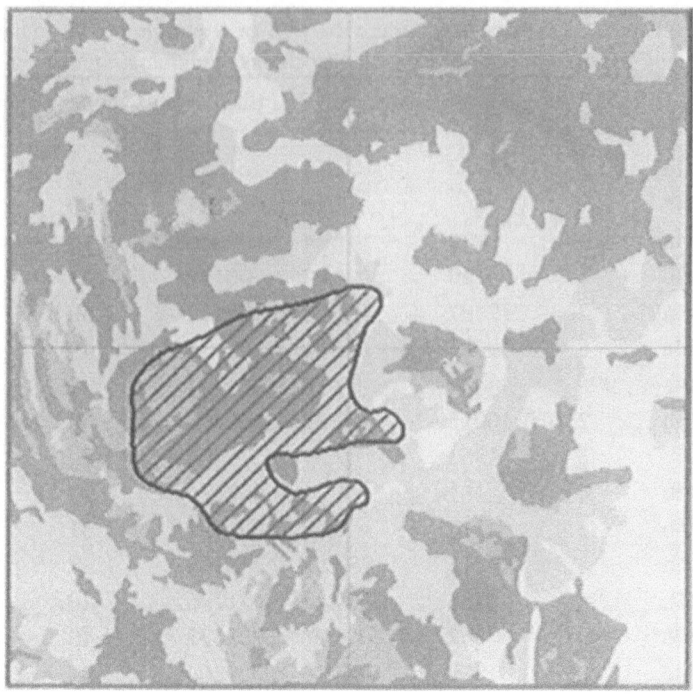

Fig. 5.7. Perimeter of "Los Barrios" large fire, affecting the natural park of Los Alcorno-cales (Andalucía, Spain). The perimeter is drawn on the top of the Behave fuel models. Potential Danger Indices computed for each fuel type were critical to obtain an accurate image of the spatial and temporal dynamism of fire danger.

and Cemagref, in France, other than the authors of this chapter. Dr. José I. Barredo and Ms. Montserrat Gómez dedicated a great effort to organise the global database and perform the logistic regression analysis. Meteorological data from the MARS project have been used to obtain the Köppen climatic map for the European Mediterranean Basin.

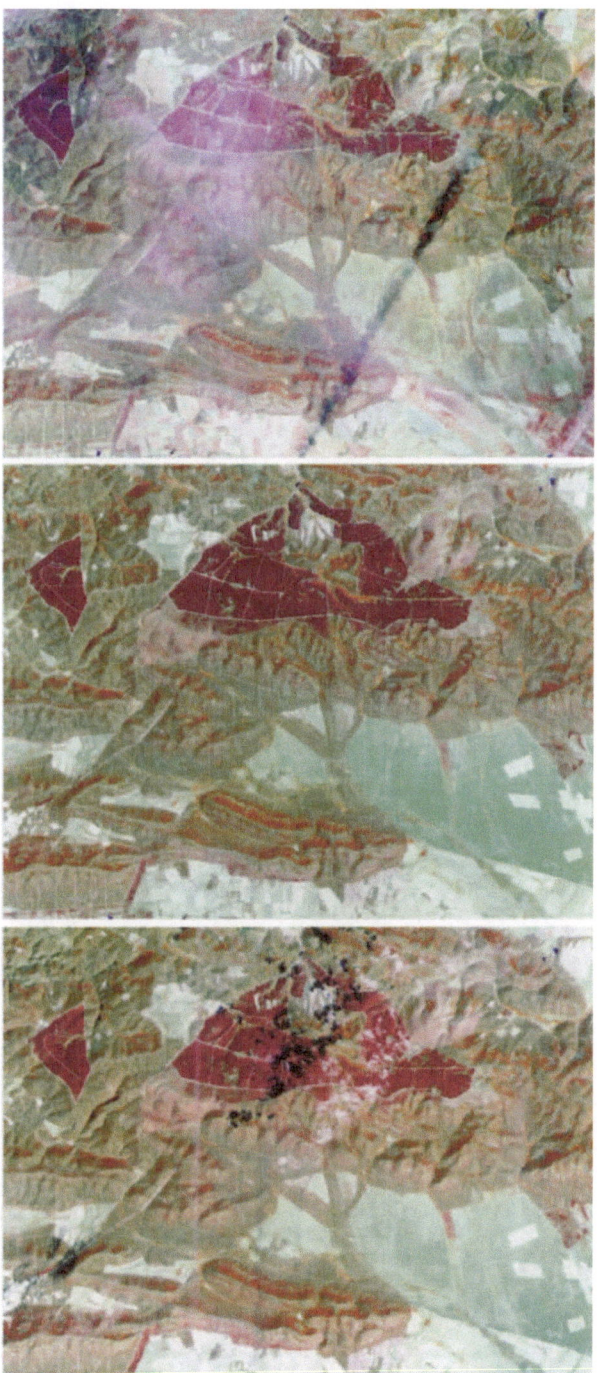

Plate 3.1. Landsat-TM images of the Cabañeros National Park: Top, April 23; Center, July 21; Bottom, September, 23.

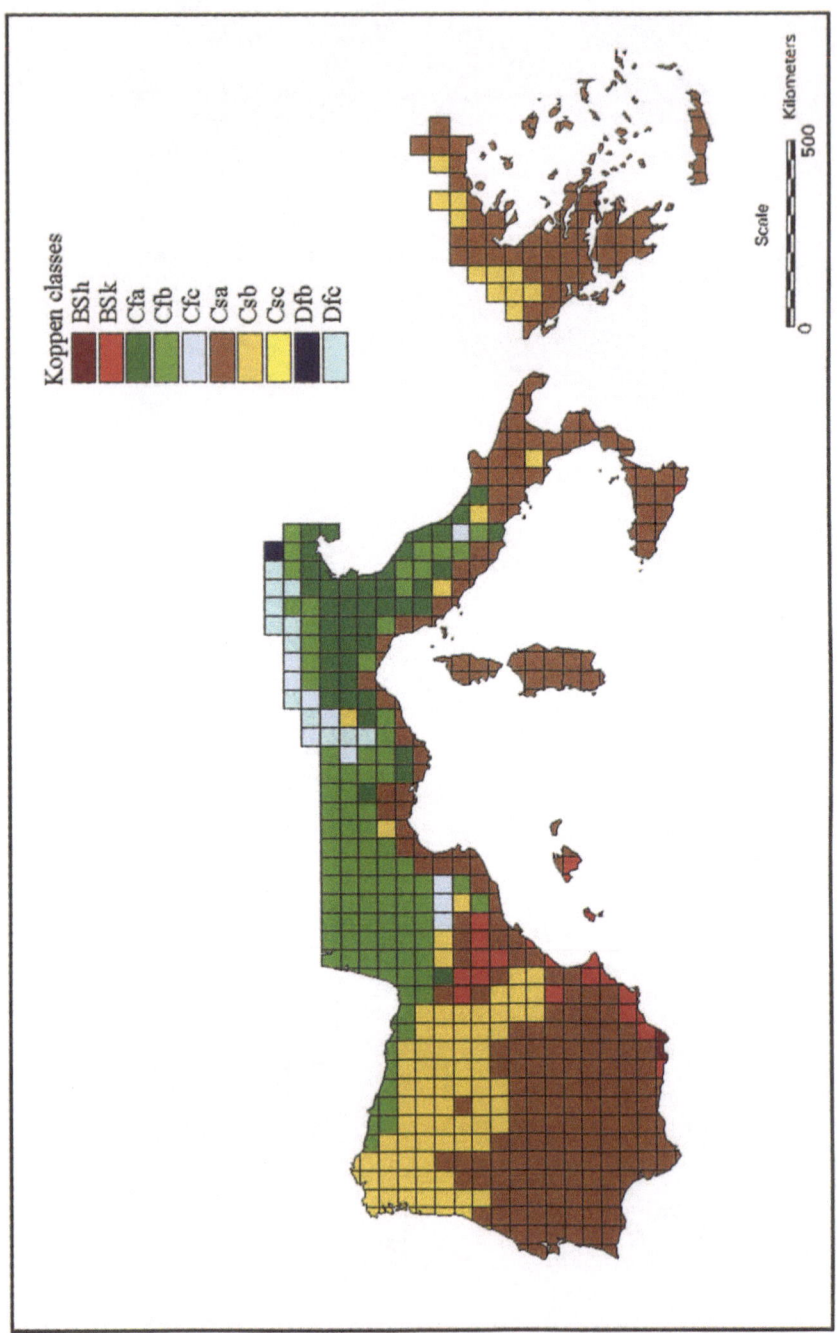

Plate 4.1 - Köppen climatic classification of the European Mediterranean basin.

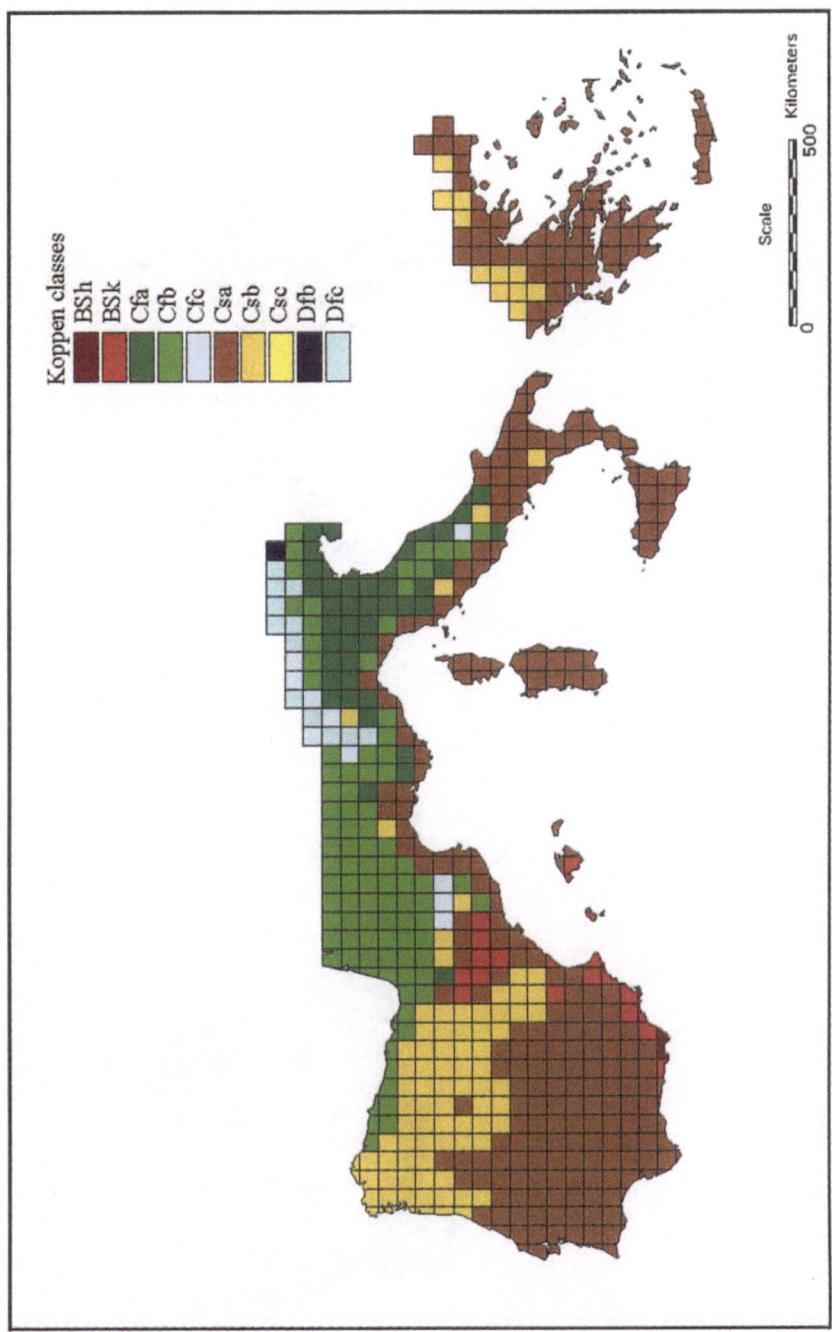

Plate 4.2. Logistic model output estimated from ST in the European Mediterranean basin (July 19-23, 1996). The mapped area is the one for which the logistic model was developed, i.e. the climatic zones Csa, Csb and BS, with urban and agricultural 50 x 50 km² grid cells masked out.

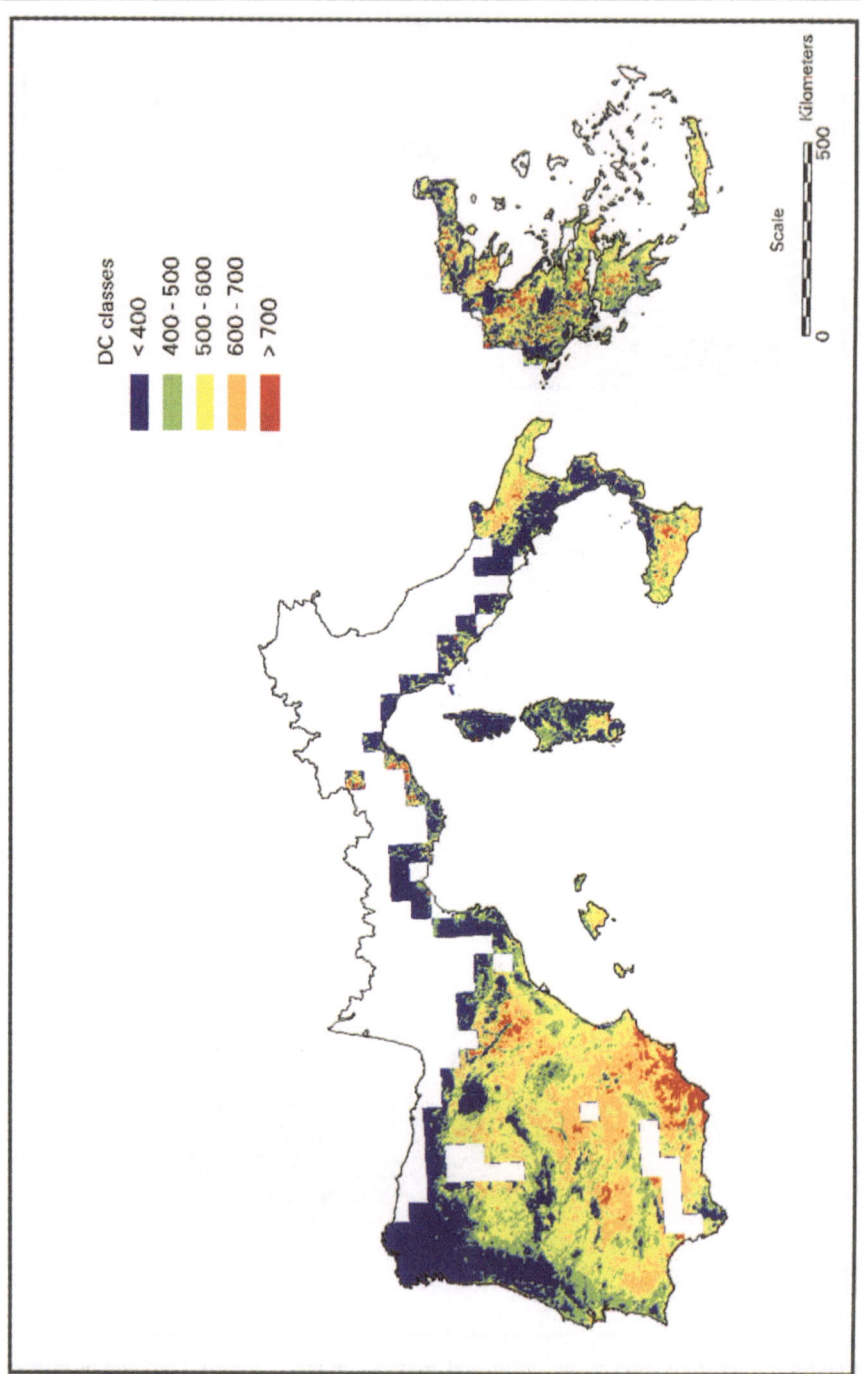

Plate 4.3. DC map estimated from NDVI in the European Mediterranean basin (July 19-23, 1996). The mapped area includes the climatic zones Csa, Csb and BS, with urban and agricultural 50 x 50 km² grid cells masked out.

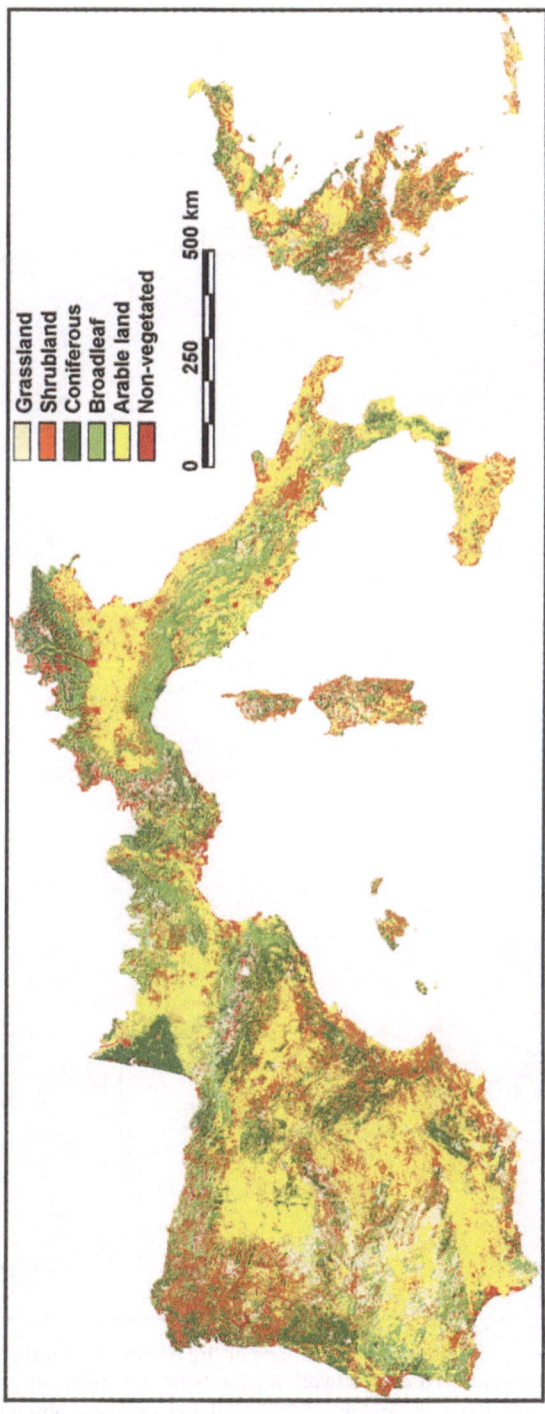

Plate 5.1. Fuel type map generated from the Corine Land Cover database.

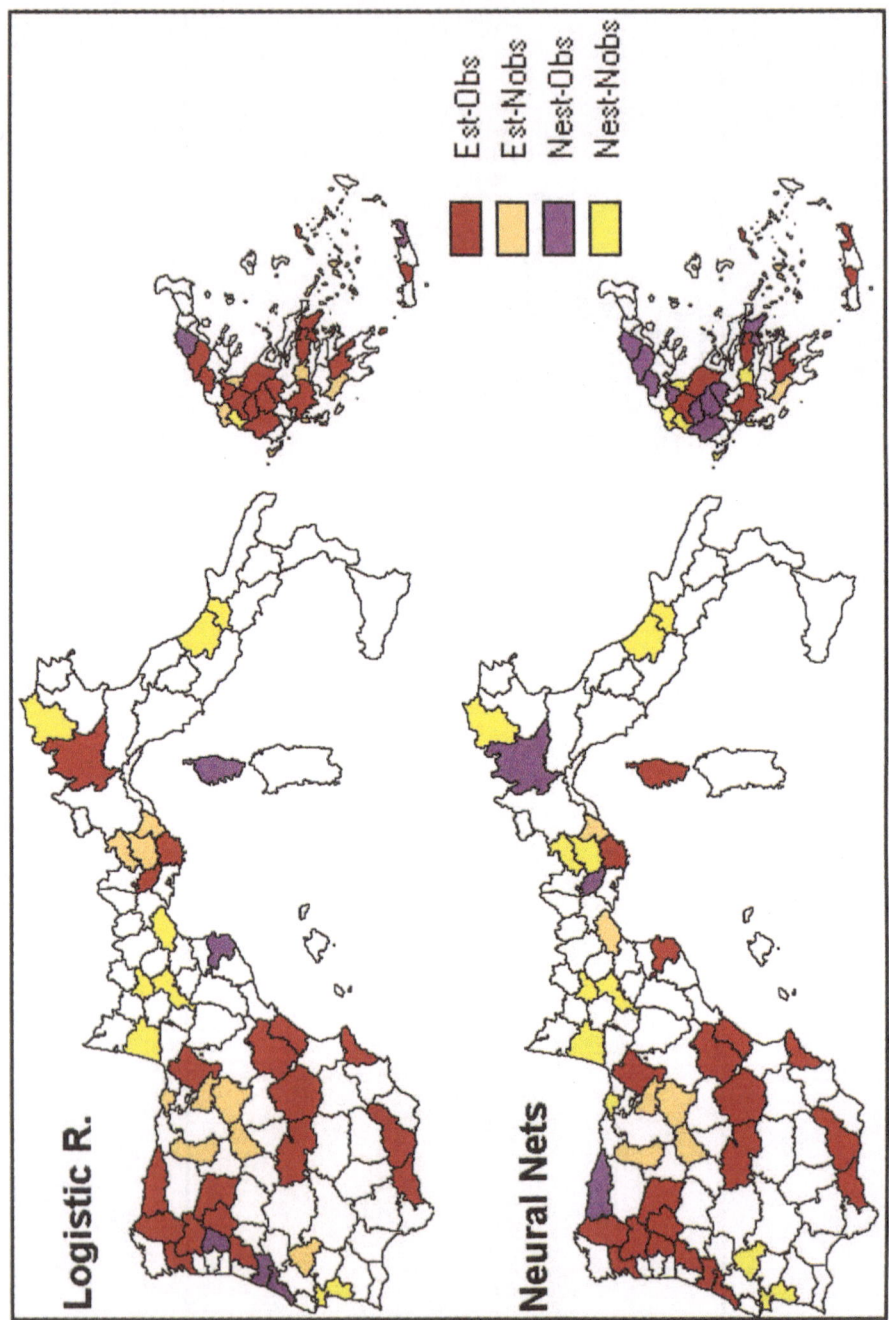

Plate 5.2. Predicted versus observed estimations of fire occurrence for the Logistic Regression model (above) and the Neural Network model (below). Est-Obs means fires that were both estimated and observed; Nest-Nobs, means both non-estimated and non-observed; Est-Nobs and Nest-Obs imply commission and omission errors, respectively.

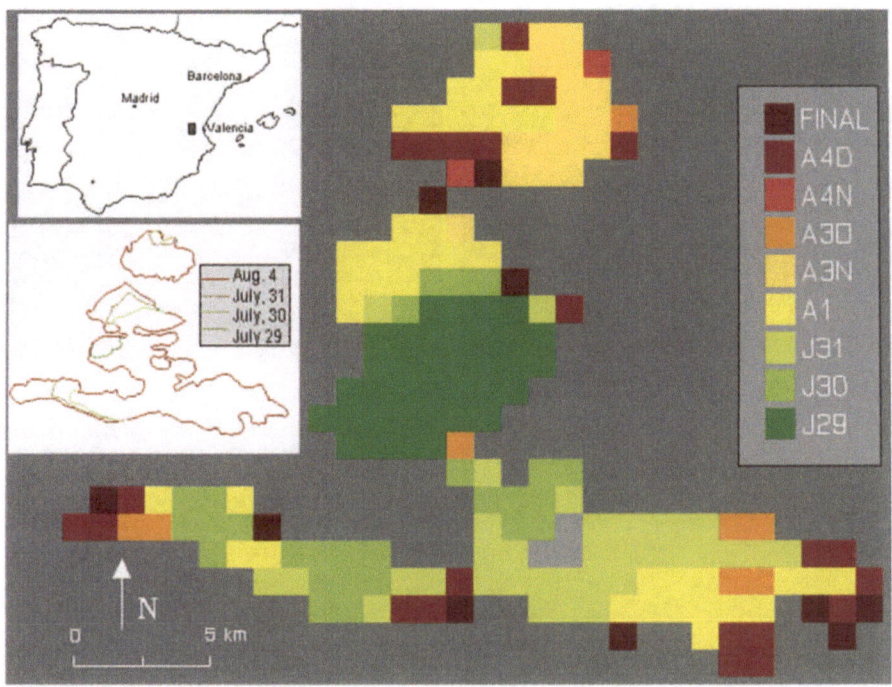

Plate 6.1. Fire growth evolution of the Buñol large fire (Valencia, Spain), derived from the analysis of NOAA-AVHRR daily data.

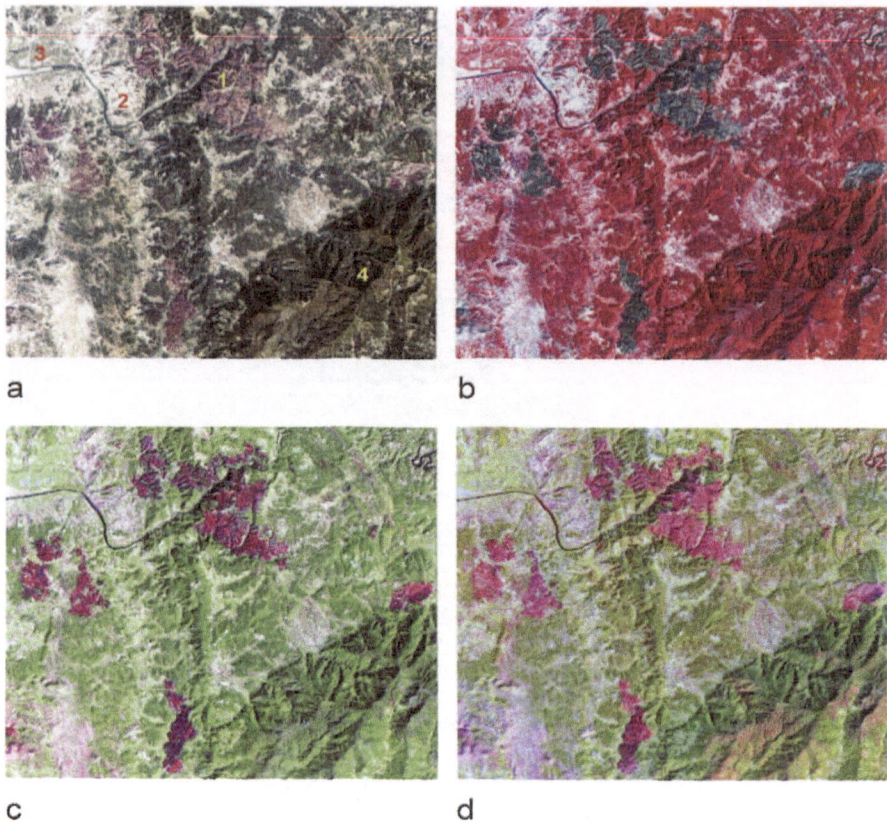

a

b

c

d

Plate 7.1. Landsat TM colour composites showing burn scars from the summer of 1995, around Coimbra, Portugal. a) RGB-321; b) RGB-432; c) RGB-743; d) RGB-647.

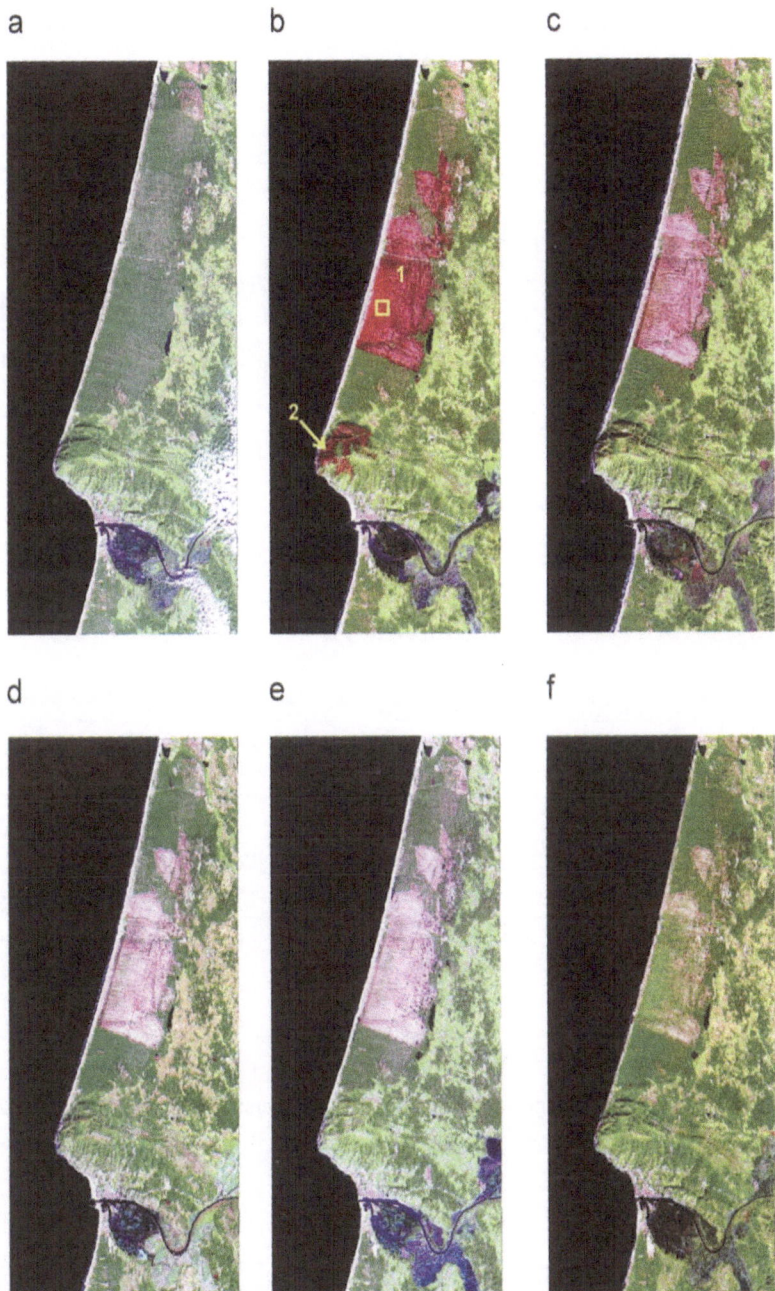

Plate 7.2. Time series of RGB-743 colour composites showing two fires from the summer of 1993, near the coast due west of Coimbra, Portugal. The images are not inter-calibrated, but were consistently histogram-stretched, to allow for visual comparison of the different dates. a) November 1992; b) October 1993; c) December 1994; d) September 1995; e) March 1996; f) October 1997. The yellow square in b) marks the data sample used in fig. 7.4.

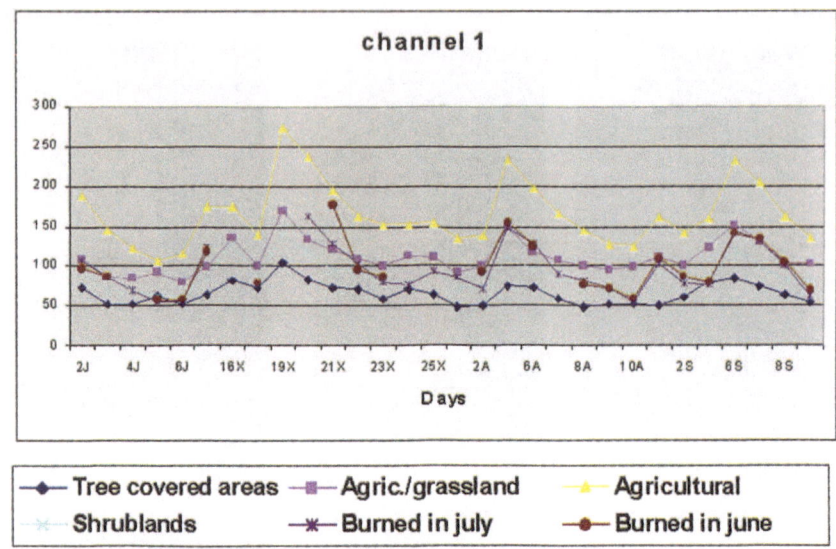

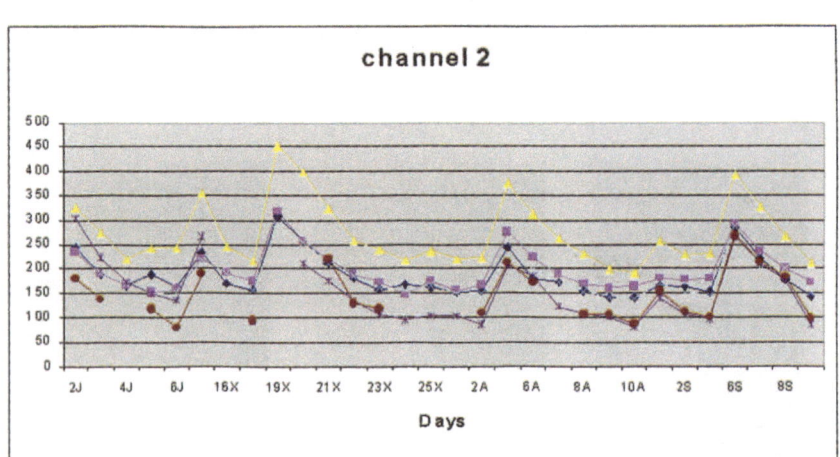

Plate 8.1. Multitemporal average values of several land covers: original channels 1 and 2

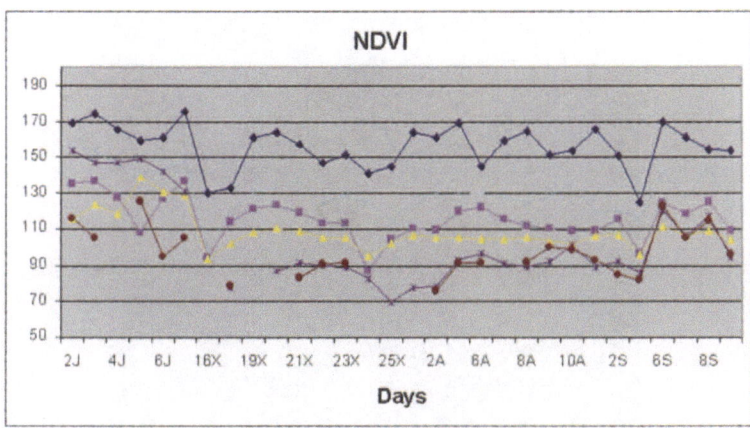

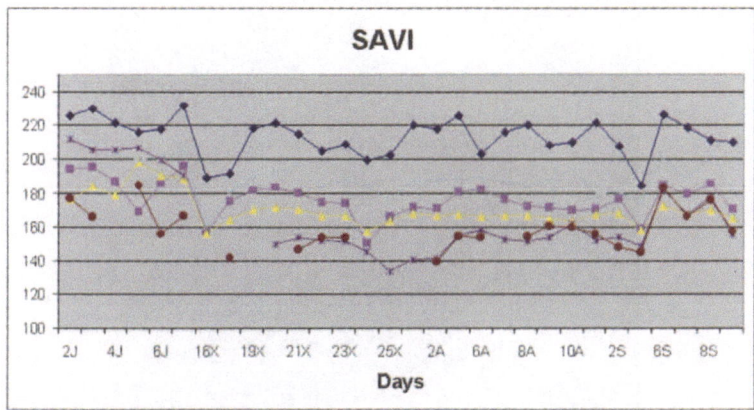

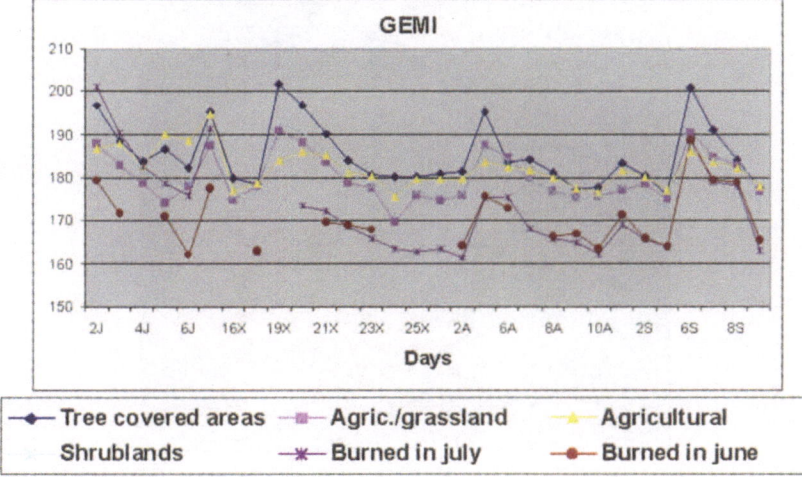

Plate 8.2. Multitemporal average values of several land covers: derived variables (NDVI, SAVI and GEMI)

Plate 8.3. Multitemporal composite from September 1991 for southern Europe, created with the minC2 → maxC4 procedure. AVHRR channels 4, 2, and 1 are in the red, green, and blue colour planes, respectively.

Plate 8.4. Multitemporal composite from September 1991 for the Iberian Peninsula. Surface temperature, GEMI, and albedo are allocated to the red, green, and blue colour planes, respectively.

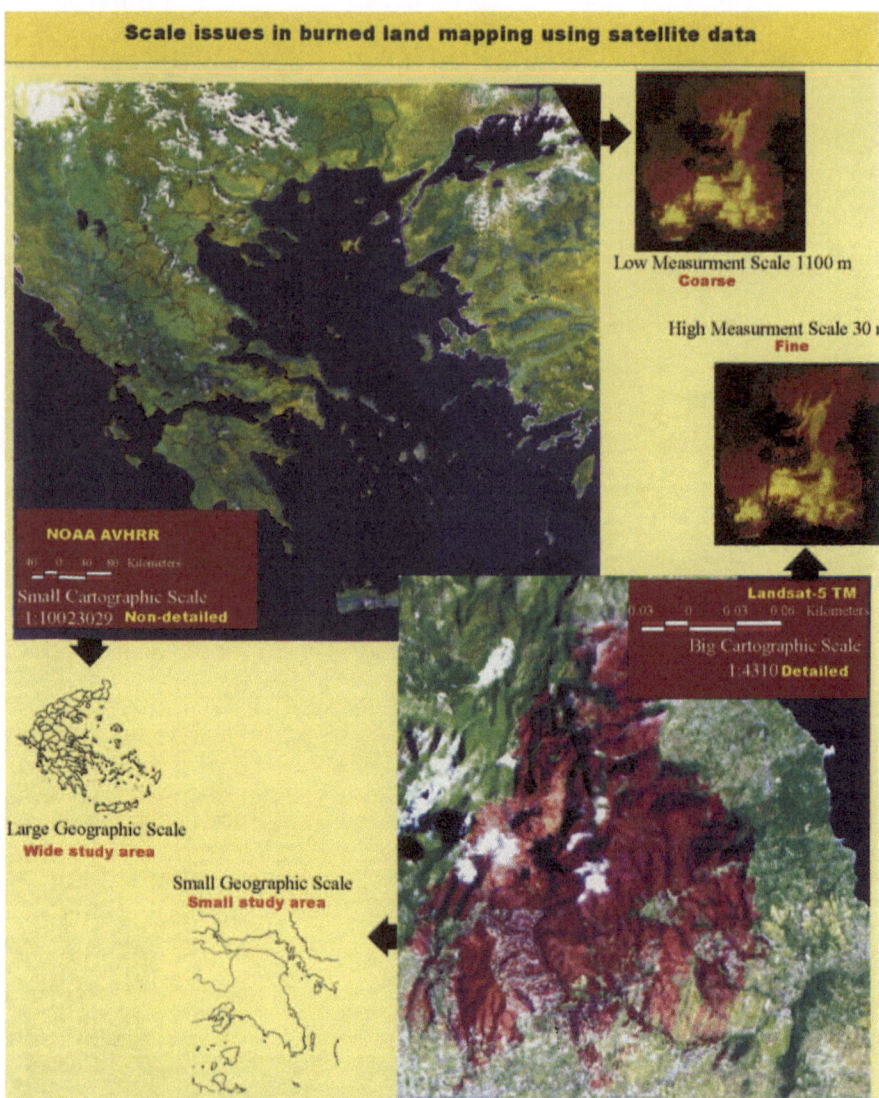

Plate 9.1. Scale is a critical component for all studies involved in the spatial or geographical domain. The four connotations of the spatial scale, that is cartographic, geographic, operational and measurement, lend to some particular characteristics of the methodological approach followed and the expected results.

Plate 9.2. One of the most particular characteristics of the Mediterranean landscapes is the high diversity and heterogeneity in terms of the spatial distribution and arrangement of the abiotic, biotic, floristic and ecological components. Sithonia peninsula, in Northeast Greece.

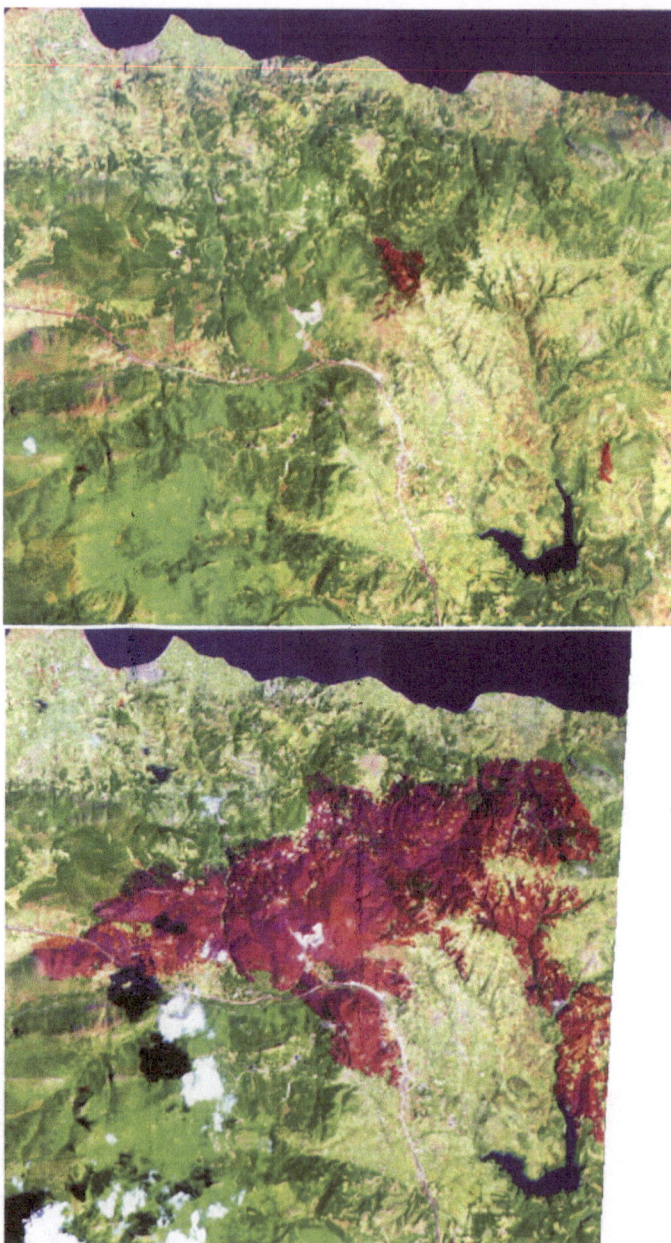

Plate 9.3. Two successive satellite images of Landsat-5 Thematic Mapper taken on 26 August (before the fire) and 11 September (just after the fire) were acquired and constituted the basic source of information for burned land mapping. The spectral channels TM7, TM4 and TM1 displayed in Red, Green and Blue color plane enhance the discrimination of the burned surfaces. Actually, the burned areas appear in dark red, while vegetative areas in green, the sea in dark blue and the bare land, clouds and urban areas in the tones of the white and pink.

6 Fire detection and fire growth monitoring using satellite data

M. Pilar Martín
Department of Geography, University of Alcalá (Spain).

Pietro Ceccato
Space Applications Institute, Joint Research Centre (Italy).

Stéphane Flasse and Ian Downey
Natural Resources Institute, The University of Greenwich (United Kingdom).

Abstract. The objective of this chapter is to review and discuss the use of near real-time satellite data for fire detection and fire growth monitoring, focusing on NOAA-AVHRR images. Capabilities and limitations of these images, as well as existing fire detection algorithms, are presented. Discussion on the potentials of future remote sensing systems for real-time fire detection concludes the chapter.

6.1 Introduction

Fire is a natural factor in the Mediterranean climate, which is characterised by high levels of vegetation stress during the summer. However, changes in traditional land use patterns have recently modified the incidence of fire in such territories, where forest fires have become more frequent and even more intense (see Chap. 2). Landscape patterns have changed over the last 30 years, increasing land cover homogeneity and as a consequence, fire risks and fire severity. Rural abandonment in the European Mediterranean countries has implied an unusual accumulation of forest fuels. According to Smith and Woodgate (1985), doubling the amount of available fuel also doubles the rate of fire spread which in turn produces a fourfold increase in fire intensity. Therefore, fire suppression becomes disproportionately more difficult as fuel accumulates. Under these conditions, large fires are likely to occur, having major effects on the productivity and stability of natural vegetation. The European Parliament has, on several occasions, identified this as a substantial regional problem and has requested both the States and the European Commission to find ways to improve the situation. This will probably have to focus on improving forest and fire management.

In this context, reduction of ecological and economical damage will always require effective fire fighting, the first essential component being the early detection of the forest fire. Traditional ground-based visual detection methods are not always appropriate for offering reliable information on fire location, size and intensity due to the small field of view one can get from the ground, often even more reduced by vegetation and/or difficult terrain. This is evident in the case of remote areas such as some boreal and tropical regions where conventional fire detection is difficult or impossible. Even if fire detection may not be considered such a critical problem in highly populated areas like the European Mediterranean Basin, where the main current detection system is based on fixed manned towers, fire-fighter personnel have underlined the importance of having accurate and frequent information on fire location and fire evolution for appropriate disaster mitigation and response. Improving the quality of current information about fire outbreaks will render the fire fighting decision-making process easier.

Remote sensing has proven to be a valuable data source in different phases of fire management both before (prevention) and after the fire (damage assessment). Methodological aspects and applications of these phases are presented in other chapters of this book. A number of authors have also demonstrated the ability of remote sensing systems to provide timely information during the fire event (detection and monitoring). This is the topic of this chapter. Remote sensing observation has significant advantages over conventional fire detection and fire monitoring methods because of its repetitive and consistent coverage over large areas of land.

There are a number of satellites and aircraft-borne remote sensing systems which can contribute to fire monitoring from space, including NOAA-AVHRR, Landsat-TM and MSS, SPOT, GOES, DMSP, ERS-ATSR, and JERS. The temporal, spectral and spatial characteristics of these instruments provide a wide range of sensing capabilities (Justice et al. 1993) and some of them have been shown to be well adapted to fire detection application. However, up to now, only NOAA-AVHRR and GOES have provided long-term operational systems, allowing low cost direct reception and near real-time fire information.

In the Mediterranean European countries, as far as fire-fighting is concerned, it is near-real time indication of a fire event at its very early stages (i.e. small fire) that will provide useful information for appropriate decision making and effective fire attack. To be useful, this information requires a synergy between temporal and spatial resolution of the satellite sensor. Current sensors were not specifically designed for such requirements, and were mostly applied initially in areas of low population density with many remote forests, such as boreal and tropical ecosystems as a tool for documenting spatial and temporal fire patterns and occurrence. However, the NOAA-AVHRR sensor is increasingly used for near real-time fire detection, and is currently being recognised as a potential useful contribution tool also in the Mediterranean areas.

6.2
Basis for fire detection from satellite data

Fire produces four forms of signal that are easily observed from space (Robinson 1991a): direct radiation from active fires (heat and light); smoke; post-fire char; and altered vegetative structure (scar). Fire detection from satellite images initially focused on analysing the first type of signal.

According to the Planck function, a black body emits spectral radiation according to its temperature, and, for a given temperature, the emission varies as a function of the wavelength. Since fires have temperatures different from other natural surfaces, observations in the appropriate wavelength should allow their detection. According to Wien's displacement law, once we know the temperature of an object, it is possible to find out the spectral region λ_{max} where the highest emitted radiance occurs:

$$\lambda_{max} = \frac{c_w}{T} \quad , \tag{1}$$

where T is the object temperature in K, and c_w is Wien's constant ($\approx$ 2898 μmK). Therefore, knowing forest fire temperature allows the selection of the most suitable spectral region for fire detection.

Forest fire temperature is difficult to estimate because it changes with, for example, fuel type, moisture content, wind, topography, and accumulated heat. Ground measurements are complex and they are strongly dependent on the location of the instruments because of the thermal anomalies that can be found within the fire. Until now, laboratory measurements, under controlled situations have been the best procedures for estimating fire temperatures for different fuel types and combustion characteristics. Using these measurements; forest fires temperatures have been estimated to lie in a range from 570 K, the minimum temperature for flaming combustion, to 1800 K, the maximum temperature in fires with heavy fuels (Robinson 1991b). According to these data and taking into consideration Wien's displacement law, the most suitable wavelength for fire detection is located in the middle infrared (MIR) spectral region between 3 and 5 μm approximately (Fig. 6.1). The radiation emitted by a fire at that wavelength being substantially higher than terrestrial background emission (Table 6.1).

While initially not designed to detect fires, most current sensors have a spectral band in this region, due to the existence of an atmospheric window located around 3.7 μm.

A number of studies on the use of remote sensing systems for forest fire detection at various scales and different regions have been undertaken in recent years (Justice et al. 1993). All of them tried to take advantage of the information obtained by different sensors in the MIR to discriminate forest fires. A brief review on radiance behaviour in this spectral region would be convenient for a better understanding of the capabilities and limitations of its use.

The MIR region presents an important contribution of solar energy, therefore, the radiance detected by the sensor is a mixture of emitted and reflected energy, simply expressed as:

$$L_{sen} = L_{ref} + L_{em} \quad , \tag{2}$$

where L_{sen} is the radiance that reaches the sensor in the MIR, L_{ref} is its solar reflected component and L_{em} is its thermal emitted component. The reflected part of the radiance reaching the sensor can be described as:

$$L_{ref} = \rho L_i T_{ref} \quad , \tag{3}$$

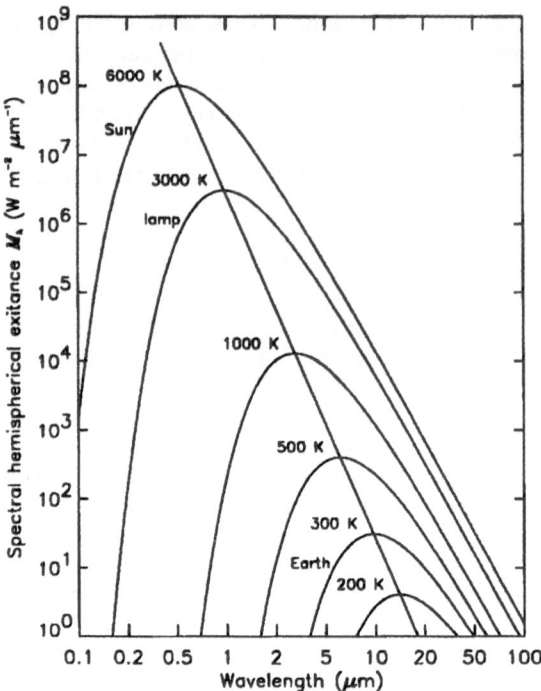

Fig. 6.1. Black body spectral hemispherical exitance according to Planck's function.

Table 6.1. Amplification factors and spectral emission maxima for background and fire temperatures at 3.7 and 10.0 μm (after Robinson 1991)

	Temper ature (K)	Peak spectral emissions (μm)	3.7 μm			10.0 μm	
			Planck radian ce from fire	Amplification over backgroud		Planck radiance from fire	Amplif. over back- ground
				No solar flux	15 % solar albedo		
Background	300	9.7	0.4	1	1	10	1
Exothermic reaction	550	5.3	146	360	130	94	9
Glowing combustion	825	3.5	1556	3900	1400	252	25
Cool forest fire	1000	2.9	3591	8900	3200	370	37
Maximum heat fire	1800	1.6	22383	55000	20000	·973	98

where:

ρ = the reflectance $\rho(\theta_i\phi_i;\ \theta_r\phi_r;\mathbf{P})$, which is a function of the geometry of illumination and observation (characterised by the solar zenith and azimuth angles, and the satellite zenith and azimuth angles, $\theta_i\phi_i$ and $\theta_r\phi_r$ respectively), and the surface parameters $\mathbf{P}$ characterising the scattering of light at that surface.

L_i = the incoming solar radiance at the surface, which is computed from $E_0\ T_i\ d^2\ \cos(\theta_i)$, where E_0 is the exoatmospheric solar irradiance, T_i is the incident beam atmospheric transmitance, d^2 is the correction coefficient for the variation in the Sun-Earth distance, and θ_i is the solar zenith angle.

T_{ref} = the reflected beam atmospheric transmitance $T_{ref}(\theta_r,\ K_{ref})$, which is a function of the satellite zenith angle θ_r, and the atmospheric attenuation coefficient K_{ref} in the solar part of the MIR.

The contribution of the emitted radiance in the MIR region is given by:

$$L_{em} = \varepsilon L(T_s)T_{em} \qquad , \qquad (4)$$

where:

$L(T_s)$ = the MIR integrated radiance computed for a surface temperature T_s following Planck's function.

ε = the emissivity of the surface.

T_{em} = the emitted beam transmitance $T_{em}(\theta_r,\ K_{em})$, which is a function of the satellite zenith angle, θ_r, and the atmospheric attenuation coefficient K_{em} in the emitted part of the MIR.

With the exception of the solar irradiance, these parameters vary strongly, depending on the time and date of acquisition (zenith and azimuth angles), the atmospheric conditions (transmitance) and the land cover type (reflectance, emissivity, temperature). In addition, one should also take into account those indirect radiances that reach the sensor (for example, the radiation from the Sun scattered into the field of view of the sensor: Cracknell 1997).

Consequently, there are a number of factors that should be taken into consideration for fire detection from space. It can be very difficult to model all the possibilities involved in the final amount of radiance that reaches the sensor. In addition, most existing MIR sensors were not designed to detect fires. The next section summarises the issues that the detection of fires from space-borne sensors must take into account.

6.3
General issues related to remote sensing of active fires

In the early 1960s, the U.S. Forest Service initiated a fire detection programme at the Northern Forest Fire Laboratory in Missoula, Montana, to develop a system capable of detecting fires in all normal atmospheric conditions, in order to improve suppression response. The result was an airborne thermal infrared line-scanning spectrometer with two spectral channels operating at 3-4 μm and 8.5-11 μm. The system reported data on forest fires (some of them had not been detected using conventional methods) with temperatures up to 600 °C, over a background with temperatures ranging from 0 to 50 °C (Hirsch et al. 1971). The possibility of using the images to provide information to fire extinction brigades was also proven, since remote sensing images enable the exact location of the fire front.

A number of case studies from different regions and using diverse image data sources have been reported since those first detection experiences using airborne sensors (Justice and Dowty 1993). However, while the potential of remote sensing images for fire detection and monitoring is significant, there are still important issues and some limitations that need to be taken into consideration.

6.3.1
Temporal issues

The usefulness of operational near real-time fire detection from space is obviously very much dependent on observation frequency. As indicated above, remote sensing images will be particularly useful if they can detect fires in their early stages. Obviously, those fires starting after image acquisition will not be detected until the next image, or missed if they are extinguished prior to the acquisition of the next one. A high frequency of observation is therefore essential if it is to make a real improvement in fire-fighting decision making.

High spatial resolution satellites such as Landsat and SPOT can contribute to fire monitoring but their cost, their centralised receiving stations and especially their low time resolution limit their use in an operational basis. Meteorological satellites are more appropriate because of their high repetition coverage. The geostationary GOES satellite series offer images every 30 minutes but only covers the American continents and cannot be used for the Mediterranean regions. Future Meteosat satellites, covering Africa and Europe, should be useful (see section on future sensors). The polar orbiting NOAA series acquire images over the same area every 12 hours for the same satellite, and cover the entire world. There are early afternoon and early morning passes. However, satellites can drift over time and at the end of their life they may overfly during the late afternoon and late morning.

Obviously, high time frequency is useful only if the data can be acquired, analysed and disseminated in near real-time. Satellites such as NOAA and GOES/Meteosat broadcast their data continuously and only require small receiving stations. A number of these stations are distributed all over the world. Local acquisition of data free of charge, analysis in situ, and fast dissemination of fire information is possible with these two satellites (Flasse et al. 1998). In mid-1993 there were over 300 AVHRR receiving stations worldwide known to NOAA (Cracknell 1997).

6.3.2
Thermal sensitivity issues

As explained above, the MIR window (3-5 μm) is particularly interesting for fire detection because it is near the spectral region for maximum radiative emissions by objects of fire temperature. Channel 3 (3.55–3.93 μm) of the AVHRR,

has been shown to be surprisingly well suited for fire detection because it is located near the peak of spectral emittance of objects at temperatures around 800 K,
i.e. the average temperature of burning grass. However channel 3 was not designed for monitoring fire events but rather for atmospheric applications (i.e.
cloud discrimination). In consequence, the MIR channel saturates at fairly low
temperatures ($\approx$ 47 °C). This saturation value can be reached not only because of
fire but also through highly reflective or hot targets, and therefore a simple threshold in this channel does not provide a sufficently accurate discrimination of fires.
Confusion with other surfaces is very common in Mediterranean areas, where
forests are often mixed with agrarian spaces or bare soil. In summer these surfaces
may reach temperatures over 50 °C (exceeding NOAA channel 3's saturation
level). In other parts of the world (for example the rain forests), this phenomenon
is not so important since fires are surrounded by dense vegetation that is relatively
cool, partly due to high evapotranspiration.

A solution to this problem is the use of night time images where temperature
and reflection resulting from the Sun are not relevant (Malingreau 1990; Langaas
1992). However, this option risks not detecting fires that are only active during the
central hours of the day (Kennedy et al. 1994), or detecting them too late. Some
other hot targets, such as certain industries and gas flares linked to oil extraction
or energy production may also saturate the MIR channel.

Confusion with highly reflective surfaces such as clouds and cloud edges is
also possible. Depending on cloud type (opacity, height, temperature and sun
zenith angle), some clouds (stratocumulus, for instance) and cloud edges can be
very reflective and saturate the MIR channel. False alarms related with saturation
of the MIR channel can also be observed due to confusion with lakes, reservoirs,
rivers and other water bodies when they present specular reflection (Belward
1991). This phenomenon, known as sun glint, only occurs for within a certain
geometry of illumination and observation.

Section 6.4 presents state-of-the-art algorithms currently available, which have
tried to take all these issues into account for the detection of fire pixel from
AVHRR observations.

6.3.3
Spatial issues

Since the objective of the near real-time remote sensing system for fire fighting
is the detection of small fires (i.e. fires at their very early stages), it is important to
realise that target fires would occupy only a small proportion of a pixel. Consequently, the radiance of a pixel will be composed of fire and background radiances. Little is known of the exact form of these contributions to the image pixel,
particularly for AVHRR. Various studies (Dowty 1993, Langaas 1995, Wooster
and Rothery 1997) underline the importance of other spatial issues in the final
pixel radiance, such as (i) the heterogeneity of the background (resulting from a
combination, for example, of burned and unburned areas), and (ii) the number,
distribution, form(s) and location of fire(s) within the pixel. Pixel size conditions
the minimum surface that has to be burning to have a signal detectable from the
satellite (Table 6.2). Belward et al. (1993) demonstrated that a bush fire with a
burning front as small as 50 m could be detected by AVHRR.

Depending on the spatial resolution, one pixel detected as a hot spot may also represent different situations. There could be one or several active fires in the area covered by the pixel, or it could be only one part of a larger fire front.

Sensors on board Landsat and SPOT provide a fine spatial resolution that would be very useful for the precise monitoring of fire growth; however, as indicated above their temporal frequency is not adequate for real-time information. While GOES-VAS provides timely observation, its spatial resolution is greater than 6 km for the thermal channels. This is certainly useful to follow the progress of huge tropical forest fires (Prins and Menzel 1994), but will limit the detection of such fires in their early stages. The NOAA-AVHRR series, although primarily designed for meteorological applications, appears to provide, amongst current sensors, the best trade-off between temporal and spatial resolution, as well as real-time acquisition.

Table 6.2. Minimum size of a fire to be detected at night times (no reflective component) in the AVHRR-MIR channel considering different observation angles and therefore different pixel sizes

A: Observation angle	2.5 times amplified background radiance (W/m² sr μm)	B: Fixed fire proportion for detectability	C: Pixel size (m²) at (A) observation angle	Fire size (m²): (B * C)
0	0.565227182	0.001116113	5207.615	5.812
5	0.552396662	0.001116113	5267.520	5.879
10	0.539566142	0.001116113	5452.360	6.085
15	0.526735622	0.001116113	5778.400	6.449
20	0.513905101	0.001116113	6275.979	7.005
25	0.501074581	0.001116113	6995.387	7.808
30	0.488244061	0.001116113	8017.649	8.949
35	0.475413541	0.001116113	9474.261	10.574
40	0.462583021	0.001116113	11584.492	12.930
45	0.449752501	0.001116113	14729.361	16.440
50	0.43692198	0.001116113	19608.141	21.885
55	0.42409146	0.001116113	27597.199	30.802
60	0.41126094	0.001116113	41660.924	46.498

Source: Chuvieco and Martín (1998).

6.3.4
Other problems related with satellite fire observation an detection

Beside all these issues mentioned above, there are additional elements that can affect satellite observation and detection of fires.

1. *Cloud and smoke*. The observation can be limited due to the existence of clouds and smoke. Even though the MIR is not as much affected by atmos-

pheric interference as other shorter wavelengths, thin clouds may prevent fire detection when fires are at cloud edge due to their size or temperature (Flannigan and Vonder Haar 1986; Hougham 1987; Setzer and Pereira 1991a). Thick clouds hide most fires, including high intensity ones. Additionally, clouds partially covering a pixel can affect fire-temperature estimation. This is one of the main limitations of the methods that attempt to assess temperatures and the area covered by the fire front. On the other hand, automatic cloud removal, which is used to prevent saturation problems in MIR channels, may lose those fires that are visible despite the presence of clouds.

As far as the presence of smoke is concerned, its influence on fire detection depends on smoke plume types (for example, depending on fire intensity, wind-fields, atmosphere stability, chemical composition and humidity content). Aerosols in the smoke plume are very effective at reducing visibility (França et al. 1995). They can generate condensing nuclei that together with water vapour may produce dense mists and hide fires (Grigoryev and Lipatov 1976).

2. *Observation geometry.* Observation geometry may be another source of error in detecting forest fires (Milne and Hall 1992). Little is known of the anisotropy of natural surfaces in the thermal spectrum. Thermal channels do not seem to be strongly affected by directional effects. Problems arise, however, when fire-affected pixels are near the edge of the image where observation angles are over 40°. Fire total intensity may not be sufficient to saturate a pixel now covering a much larger area. Pixel overlap can also reach almost 100% at the very edge. Consequently, it may happen that an active fire that should only be detected by one pixel is detected simultaneously by several pixels at the image edge (Stezer and Pereira 1991b).

3. *Small underlayer fires.* Dense forest may prevent detecting small underlayer or peat fires.

Detection of fires depends on fire characteristics and environmental conditions but also on sensor noise and its degradation, background temperature and atmospheric contamination. Therefore, it is extremely difficult to find a detection algorithm valid both in space and time and to apply a consistent methodology for automatic fire detection from space at regional or global scales. In trying to solve this problem, sophisticated detection algorithms have been developed. Most of them will be reviewed in the next section.

6.4
Active fire detection with NOAA-AVHRR images

The characteristics and issues presented have led several authors to develop algorithms for fire detection, most of them using NOAA-AVHRR images. Robinson (1991b), Kennedy (1992), and Justice and Dowty (1993) provide thorough discussions of fire detection algorithms and cite case studies of their application. This section completes these with summaries of recent developments.

Fire detection is achieved by one of the four methods described below, all of which use the dominating effect of hot fires in Channel 3 signals:

- Channel 3 single threshold algorithms.
- Multi-channel threshold algorithms.

- Contextual algorithms.
- Sub-pixel fire detection algorithms.

6.4.1
Channel 3 single threshold algorithms

The Channel 3 single threshold approach relies on the assumption that only the Channel 3 data are required and that a single (empirically derived) threshold value can be used to identify pixels with sub-resolution fires (Malingreau and Tucker 1988, Setzer and Pereira 1991b, and Pereira and Setzer 1993). The simplest algorithm is to retain all pixels which are saturated (or near saturated) in Channel 3. Using this approach, Muirhead and Cracknell (1985) were probably the first to report fire detection in the European regions using NOAA-AVHRR data. A similar approach has been applied in some geographical regions with intense biomass burning (the Amazon Basin in Brazil), where the saturation level of Channel 3 has been considered appropriate and is being used for operational detection (Setzer and Pereira 1991b). In other regions, Channel 3 values, just below saturation, have been considered more suitable (Kaufman et al. 1992, Matson and Holben 1987).

However, many authors have found that this simple threshold can lead to false detection due to variations in environmental conditions and AVHRR instrument response (Frederiksen et al.1990, Grégoire et al. 1993, Setzer and Malingreau 1993, Setzer and Verstraete 1994). They arrive at the conclusion that different thresholds need to be applied at night and during the day (i.e. at different overpass times), for different ecosystems, and for different seasons. Added to this, simple Channel 3 thresholding is susceptible to confusion with hot bare surfaces and bright clouds, as discussed earlier, and may not be suitable for regional scale fire studies over time. There is considerable work from West Africa covering this issue (Grégoire et al. 1993). This problem is not so acute in regions where surfaces are relatively cooler and which exhibit less seasonal land surface variation (for example, equatorial forests of Congo and Amazon).

A solution to this problem relies on establishing multiple thresholds combining information from Channel 3 and from other thermal channels. This approach is described in the next section.

6.4.2
Multi-channel threshold algorithms

Multi-channel threshold algorithms have gained a great deal of support as they have been shown to be regionally robust and simple to implement. They are seen as an improvement on single channel threshold algorithms, because of the extra spectral data taken into account to reduce false detection, and of the easiness of implementation of a set of decision rules (Kennedy et al. 1994).

These algorithms use combinations of two or more fixed thresholds for Channel 3 and 4 (3.75 μm and 10.8 μm, respectively) alone or in combination. They are usually based on calculated brightness temperatures T^B and operate either on a pixel-by-pixel or a contextual (pixel-by-neighbourhood) basis. This latter approach (also called the spatial technique; Justice and Dowty 1993) is discussed in Section 4.3 below.

In the pixel-by-pixel multi-channel thresholding approach a series of tests are employed, all of which must be fulfilled for a pixel to be classified as a fire. These multi-channel algorithms usually take the following form:

1. $T^B(3) > k_1$

2. $T^B(3) - T^B(4) > k_2$

3. $T^B(4) > k_3$,

where $T^B(3)$ represents the brightness temperature for Channel 3, $T^B(4)$ represents the brightness temperature for Channel 4, and k_i is the brightness temperature threshold for test i=1,2,3.

Similarly to the single threshold algorithm, the first test of the multi-threshold approach looks for high Channel 3 brightness temperature, i.e. pixel affected by fire. However, as explained above, this will also select other pixels such as hot soils. Therefore the second criterion seeks to discriminate between fire pixels and non-fire but warm surface pixels. This test is based on the contrast between the middle and thermal infrared response at different temperatures. As described above, an object at fire temperature will emit much more radiance in the MIR than at other wavelengths. Consequently, a pixel containing a fire (temperature over 500 K) will show higher brightness temperature in Channel 3 than in Channel 4, compared to a pixel containing solely hot bare soil (temperature not higher than 330 K). Consequently, the difference between $T^B(3)$ and $T^B(4)$ will be higher for a pixel affected by a fire. Recent observations have indicated that in the few cases where a pixel is largely covered by an active fire, the wide wavelength range Channel 4 receives enough radiance to increase $T^B(4)$ as far as saturation. This test $T^B(3) - T^B(4) > k_2$ will miss such cases. The third criterion is a crude cloud check to reduce the possibility of false detection due to highly reflective clouds that may also saturate Channel 3. In addition to these thermal channels tests, other tests on reflectance channels are often used to diminish false detection due to highly reflective pixels caused by clouds and sun glint.

As in the case of simple Channel 3 thresholding, all the published multi-channel thresholds have been developed for specific regions and are based on empirical approaches. A wide range of multi-channel threshold criteria exist (Kaufman et al. 1989, and 1990a,b; Langaas 1989, 1992 and 1993; Belward et al. 1994; Kennedy et al. 1994 and França et al. 1995).

The approach of Kaufman et al. (1990a,b) uses AVHRR channels 3 and 4 and was developed for forest environments in Brazil. Fire detection is triggered by satisfying three criteria:

$T^B(3) \geq 316$ K;

$T^B(3) - T^B(4) \geq 10$ K;

$T^B(4) > 250$ K.

To adapt the proposed thresholds to other specific areas, such as savannah in West Africa, some modifications to the Kaufman methodology have been suggested (Kennedy et al. 1994). Land surfaces in savannah regions are generally much warmer than in the Brazilian forests. Consequently, a temperature difference of 10 K between Channel 3 and Channel 4 is too small to eliminate the warm, non-fire surfaces. Modifications also seek to avoid high temperature Channel 3 pixels caused by highly reflective clouds that could also be selected by the Kaufman test because of high radiant temperatures in Channel 3 and low in Channel 4. Moreover, Kennedy et al. (1994) found that many large areas in the northern parts of the Sudanian and Sahelian zones were included as spurious fires early in the year. They concluded that this confusion could be caused by highly reflective bare soils that can cause Channel 3 to saturate and include an extra test based on 'top-of-atmosphere' reflectance data from Channel 2, where fire and recently burned areas will tend to have low reflectance. The revised approach took the following form:

$T^B(3) > 320$ K;
$T^B(4) > 295$ K;
$T^B(3) - T^B(4) > 15$ K;
$\rho_2 \leq 16\%$,

where ρ_2 is the Channel 2 top-of-atmosphere reflectance. This method has been proposed as one of the most appropriate automated approaches to detect fires throughout the ecosystems of West Africa, although it is clear that the effect of highly reflective Sahelian surfaces is extraordinarily variable year to year and difficult to account for consistently (Kennedy et al. 1994).

França et al. (1995) have extended the multi-threshold algorithm to include two more tests. The first one $[0 \leq T^B(4) - T^B(5) \leq 5$ K] allows a separation of different surfaces, especially pixels partially covered by clouds. The second test (equivalent albedo in Channel 1 < 9%) is applied to exclude pixels where atmosphere with dry haze or aerosol concentrations may cause discrimination problems.

Koffi et al. (1995) applied the same methodology developed by Kennedy but adapted the thresholds according to ecological regions and time period. Threshold values have been selected using statistical analyses of visually observed fire pixels (associated with smoke plumes in Channel 1).

The European Space Agency (ESA)/ESRIN uses a multi-threshold algorithm to generate continental fire products. It has been documented by Arino et al. (1993) and uses the following tests:

$T^B(3) > 320$ K;
$T^B(4) > 245$ K;
$T^B(3) - T^B(4) > 15$ K;
$\rho_1 < 25\%$ (prevent false detection due to highly reflective surfaces)
$|\rho_1 . \rho_2| > 0.01$ (prevent false detection due to Sun glint),

where ρ_1 and ρ_2 and Channel 1 and 2 top-of-atmosphere reflectance respectively.

If at least one of these conditions is not satisfied, the pixel is classified as non-fire. In practice a sixth step based on visual inspection is employed to eliminate pixels in which obvious false detection occurs but which have passed previous tests.

The main disadvantage of pixel-by-pixel absolute thresholding approaches is that they are insensitive to variations in normal land surface temperature conditions over the season for a given area, and require a lot of spatial and temporal adjustments.

6.4.3
Contextual algorithms

The multi-channel threshold algorithms will work very well as long as thresholds are appropriate for a particular ecosystem, at a specific time. This is actually the main constraint of absolute threshold algorithms when automated detection is concerned. This has led recent developments in fire detection algorithms to focus on ways of adapting thresholds automatically, hence leading to contextual algorithms (also called spatial analysis techniques). The contextual approach is based on the same idea of a hierarchy of T^B test as described above, but automatically updating thresholds for each pixel using the neighbouring pixels to the target pixel as an indicator of local environment radiometric characteristics, hence taking into account spatial and temporal evolution of the local background signal.

In 1986, Flannigan and Vonder Haar already included the contextual approach in their technique, but for only two tests (on Channels 3 and 4). Remaining tests (differences between Channels 3 and 4) still operating with absolute thresholds. Lee and Tag (1990) also included a contextual approach, where neighbouring pixels information is used to 'correct' the fire target pixel which is then compared to sub-pixel resolution hot spot temperature (this approach was only developed for night time images). In 1991, Smith and Vaughan adopted a contextual approach based on three tests of Channel 3, two of them using contextual information. A 5x5 pixel window centred on a candidate fire pixel is used to compute mean and standard deviation of Channel 3 surrounding pixel values. The standard deviation is used to identify noise, and the mean value is compared with the candidate pixel value and if the difference is larger than a certain threshold, the candidate pixel is identified as fire pixel.

More sophisticated contextual algorithms have been developed lately, using most of the information available from AVHRR spectral channels plus statistical information to create, automatically, individual pixel thresholds. These algorithms are based on a two-step approach:

- *Potential fire pixel selection:* to select candidate pixels that could potentially be fires, using simple absolute thresholds (this step is not essential, but it considerably reduces the calculation);
- *Fire confirmation*: confirm that a potential fire pixel contains an active fire, by comparing its brightness temperature T^B with its surrounding ones (background/context).

In 1993, Justice and Dowty reported the adaptation of a contextual algorithm used for fire detection with GOES images (Prins and Menzel, 1994). Flasse and Ceccato (1996) extended the approach, in order to take into account the large variety of fires in different ecosystems and allow automated detection of fires for global application. These algorithms are presented in Table 6.3.

The multi-threshold algorithm of Koffi (1995), presented in Section 6.4.2, and the contextual algorithm of Flasse and Ceccato were applied to the wide range of ecosystems of Central African Republic where they provided very similar results. However, while the former required a labour-intensive definition of fire detection thresholds for each ecosystem and month, the latter detected fires automatically and substantially faster (Eva and Flasse 1996).

Table 6.3. Summary of criteria used in fire detection from NOAA-AVHRR data

	Justice and Dowty (1993)	Flasse and Ceccato (1996)
Step A: *Potential fire pixel selection*	$T^B(3) \geq 316$ K $T^B(4) \geq 290$ K $T^B(3) > T^B(4)$	$T^B(3) > 311$ K $T^B(3) - T^B(4) > 8$ K $\rho_2 < 20\%$ No $T^B(4)$ test is applied since a cloud mask is applied before the algorithm is run
Step B: *Confirmation*		
Fire-background pixels	Fire-background pixels are all pixels within the context window except those selected as potential fire (PF) pixels.	Fire-background pixels are all pixels within the context-window except those selected as PF pixels or masked as cloud, water body or desert.
Context-Window	A window varying from 3x3 to 21x21 pixels around the potential fire to operate until there are 3 or more fire-background pixels, occupying at least 25% of the window.	A window varying from 3x3 to 15x15 pixels around the potential fire to operate until there are 3 or more fire-background pixels, occupying at least 25% of the window
Statistics computation	Statistics are only computed for fire-background pixels.	Statistics are only computed for fire-background pixels.
Decision	A PF is confirmed as fire if $T^B(3\text{-}4)_{PF} >$ $\max\{[T^B(3\text{-}4)+2\sigma]_{BM}; 3\}$ Where PF stands for potential fire, BM for background mean, and σ for mean standard deviation. If conditions for a context-window were not met with the maximum size window, a PF is NOT considered as fire.	A PF is confirmed as fire when both following tests are satisfied: $T^B(3\text{-}4)_{PF} > [T^B(3\text{-}4) + 2\sigma]_{BM}$ $T^B(3) - [T^B(3) + 2\sigma]_{BM} > 3$ K If conditions for a context-window were not met with the maximum size window, a PF is NOT considered as fire.

Giglio et al. (1998) recently provided a good performance comparison of the algorithms of Justice and Dowty, Flasse and Ceccato, and Arino et al. (1993), clearly indicating the power of the contextual approach. The IGBP-DIS Global Fire Products based its fire detection algorithm on the above contextual approaches to produce world maps of fires (see Internet address: http://www.mtv.sai.jrc.it/projects/fire/gfp).

In the past few years, the Flasse and Ceccato algorithm has been extensively used in the framework of development projects in many different areas of the tropics (Flasse et al. 1997, 1998) allowing the application and testing of the approach into many different regions and seasons. These recent experiences have led the authors to improve the algorithm in order to take into account a larger number of possible cases. Recent improvements are summarised here:

1. The performance of the algorithm gets better when the minimum context-window is set to 5x5, to start with a better sampling of the background (another way would be to increase the minimum number of background pixels for statistics computation).

2. A background distribution criterion has been included to prevent false detection due to edges (for example, hot soils at the edge of a desert will be detected as PF, and falsely confirmed as fire because of the relatively cool edge of the desert. Similar false detection is found at the edge of water bodies and clouds, when these water or cloud pixels have not been picked up by masking processes). The background distribution test consists of dividing the context window into four regions, and in rejecting statistics computation if there is not at least one fire-background pixel in each of at least three regions.

3. A new test has been included to allow the detection of very hot and/or very large fire fronts which provide enough radiation to substantially increase $T^B(4)$. Such fires are quite common in Southern Africa savannah, and are usually missed when the traditional $T^B(3)$ - $T^B(4)$ test is used. However, the latter was kept as a part of the PF selection because substantially reducing the number of PF (such as those warm soils detected as PF), and a new test was introduced: a pixel will also be selected as PF when $T^B(4)_{PF} > T^B(4)_{BM}$. Research is currently under way to find a generic way to take all fire sizes and intensities into account.

4. Finally, while the algorithm already provides very reliable results when applied globally, it has been noted that its performance improves when the absolute thresholds for potential fire selection are tuned to a particular country or region. This can diminish the risk for false detection and increase the processing speed. For example, in sahelian regions, it is rare that brightness temperatures for Channel 3 are lower than 315 K. In some regions such as in Spain, recent research has shown that some fires can have brightness temperatures for channel 3 lower than 311 K. It is necessary in that case to lower the threshold value in Channel 3 used for identifying potential fire pixels.

Harris (1996) suggested that one should consider as fire pixels those pixels presenting a local variation that is higher than 'natural variation' within the image. In other words, the technique tests the local difference between the target pixel and its neighbouring pixels against the maximum difference that can be found, in an area without any fires, between pixels and their surroundings (the natural variation). Since every single pixel of an image is tested, i.e. there is no pre-filter to detect potential fires, fires in the immediate neighbourhood of target pixel may affect the statistics so that the target pixel is not detected as fire. Therefore the technique excludes from the computation any new pixel detected as fire, and iterates until no new fire pixels are detected.

In Mediterranean Europe, Lasaponara et al. (1998) suggest an improved fire detection algorithm using AVHRR images acquired over Italy. They suggest adding a temporal context to the spatial contextual approach. It consists in comparing 'today's' pixel with the mean of the last 10 days for a spatial box ranging from 3x3 to 7x7 pixels. This appears to be also very promising for fully automated detection.

One of the work packages within the MEGAFiReS project dealt with forest fire detection using AVHRR data in Mediterranean Europe. The main task of this package was to develop an algorithm able to search automatically for fire-affected pixels in order to test the efficiency of low spatial resolution satellite sensors such as NOAA-AVHRR for improving fire detection and fire fighting operations in Mediterranean forest. Based on the literature, different multi-threshold and contextual algorithms were tested but no definitive conclusions could be reached due to the lack of validation data (Infocarto, 1998).

However, quite clearly these recent developments of contextual approaches have opened new doors which will facilitate the near future use of operational and automated algorithms with increased reliability.

6.4.4
Sub-pixel fire detection algorithm

As was mentioned in the introduction to this chapter, effective fire detection from satellite data in Mediterranean areas requires location of a fire in its early stages: this generally means smaller pixel size. Therefore it would be useful to gather information on fire size at a sub-pixel level.

Sub-pixel fire detection consists of decomposing the energy received by the sensor into fire and background energy. However, as reviewed in previous sections, the application of this approach is currently limited with the low sensitivity of current sensors. Indeed, fires can be evaluated with this approach only if their contribution to the containing pixel's radiant flux is sufficient to permit detection but insufficient to cause saturation. Guidance on the suitability of these algorithms is given by Robinson (1991a,b) and by Kennedy (1992). We briefly describe here one of the most widely used sub-pixel algorithms developed by Matson and Dozier (1981). This is a bispectral deterministic invertible model, which utilises the non-linear nature of the Planck function to link the brightness temperature of an object to the emitted energy that it radiates. It has been shown that it is possible to calculate sub-resolution brightness temperatures and the fractions of pixels occupied by fires, by solving two simultaneous equations (Dozier 1981) of the form:

$$L_i(T_i) = pL_i(T_f) + (1-p)L_i(T_b), \quad i = 3,4 \qquad , \tag{5}$$

where $L_i(T_i)$ is Channel-i radiance as a function of the pixel temperature T_i, $L_i(T_f)$ and $L_i(T_b)$ are the portions of Channel-i radiance as a function of the fire temperature T_f and background temperature T_b respectively, and p is percentage of the pixel area covered by the active fire.

The technique allows the calculation of two sub-pixel areas (fire zone and background) within an unsaturated pixel. By comparing a specified fire threshold temperature with the calculated fire temperature, it is possible to identify a fire in the pixel. Background temperature should be specified previously and this can be done in two ways:

- Using a look-up table of background temperatures related to the zone of interest.
- Calculating background temperatures using the responses of surrounding pixels in Channels 3, 4 and 5.

A different fire target temperature is used, and defined with local experience of fire behaviour.

The flux to the sensor varies with the respective sizes and temperatures of fire and background. For a given background temperature, an envelope of resolvable fire events can be plotted as a function of size and temperature.

The Dozier model is constrained, however, by some of its unrealistic assumptions (Langaas and Muirhead 1989). These are:

- any scene (pixel) has only two uniform classes (high temperature fire and low temperature background),
- all fires within a pixel (1 km²) are grouped together,
- both fire and background have black body emissivity properties,
- the background has spatially homogenous thermal properties.

The model has been restricted to a number of short duration case studies (Matson and Dozier 1981, Flannigan and Vonder Haar 1986 and Illera et al. 1995) largely because of the difficulties in automating a system to derive local background temperatures from sampled neighbouring pixels. Lee and Tag (1990) attempted to develop a means of parameterising background temperatures to remove this difficulty and enable fire studies regionally and over longer periods.

6.4.5
Additional issues

Even though the algorithms discussed in this chapter have been developed taking care to include as many detection situations as possible, they are not perfect. In some cases, false detection still happens. As explained above, AVHRR-Channel 3 produces fire-like response when bright clouds, hot soils, and glint over water bodies are present in the image. As well as the reviewed algorithms, other sources of available information should be used to secure reliable fire products. It is important for example to mask out all areas that cannot take fire, such as:

Water bodies. Good results were obtained with a simple technique consisting in building a water mask by keeping all NDVI pixel < -0.12 over a representative period of time (Flasse and Ceccato 1996). There is also global land cover information available such as the World Data Bank II (WDB II, 1991,U.S. Department of Commerce), or the World Vector Shoreline (WVS) that can be used to build water mask.

Deserts. Tucker et al. (1994) use AVHRR data to assess desert area. The Olson World Ecosystem Classes can also be used, and is contained within the Global Ecosystems Database published by NOAA.

Clouds. While care can be taken to avoid a cloud being mistaken as a fire (see thresholds algorithms above), it is still quite interesting to assess the area covered by clouds, i.e. where fire is not identified from the images but can be by alternative methods based on aerial or ground observation. There are a variety of cloud detection approaches. Experience led to suggesting a simple approach for the Global Vegetation Fire Product (Stuttard et al. 1995), which has been shown to be quite efficient in most areas of the world (given some local adjustments). Using this algorithm a pixel is labelled as being a cloud when it satisfies all tests included in Table 6.4.

Table 6.4. Some criterion for cloud-masking

DAY time	NIGHT time
$\rho_1 + \rho_2 > 120\%$	
$T^B(5) \ < 265\ K$	$T^B(5) \quad <285\ K$
$\rho_1 + \rho_2 > 85\%$ AND $T^B(5) < 285\ K$	
$\rho_1 > \rho_2$ AND $\rho_1 + \rho_2 > 50\%$	
Where ρ_i is top-of-atmosphere reflectance for Channel 1 and 2, and $T^B(5)$ is the brightness temperature for Channel 5.	

Glint. As indicated above, Sun glint is the result of the specular reflection of light on water bodies and often saturates Channel 3. It is characterised by specific geometry of illumination and observation, and simple tests on satellite and Sun angles should suffice. Arino et al. (1993) also suggested a glint test based on Channel 1 and 2 reflectance (see above).

6.5
Fire growth monitoring using AVHRR images

Temporal resolution of AVHRR data may also be used to follow the spatial evolution of large fires, providing significant information for fire behaviour modelling.

Fire managers require spatial information on fire evolution in order to plan fire fighting activities better. In recent years, a large number of projects have dealt with the prediction of spatial fire behaviour, by coupling fire modelling programs and Geographic Information Systems (Campbell 1995; Holder and Wyngaarden 1990; Pereira and Vasconcelos 1990; Vasconcelos and Guertin 1992). One of the main difficulties of this type of research project is a lack of validation of field information on the spatial progression of the fire, either because of poor visibility from the field, or concentration of aerial resources on fire suppression while the fire is active.

Within this context AVHRR images can provide valuable information because of the possibility of monitoring fire growth at least every 6 hours (when using two NOAA satellites, morning and afternoon). Coarse spatial resolution of AVHRR data restrict this potential to large fires, whose size and duration are enough to be followed in time series of AVHRR image data.

This potential has been explored in several studies in the Mediterranean (Chuvieco and Martín 1994; Pozo et al. 1997). The methods proposed by these authors

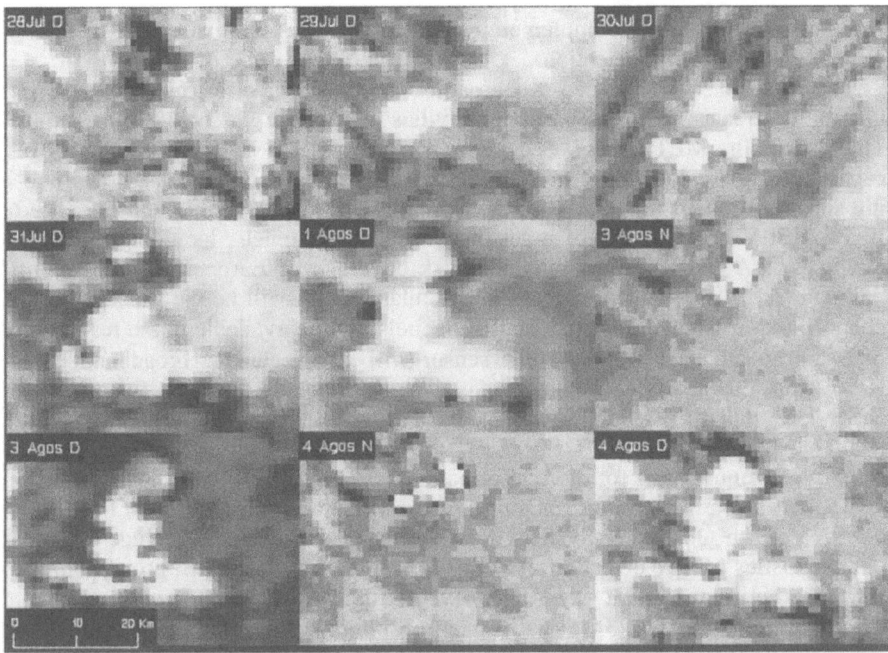

Fig. 6.2. Spatial evolution of the Buñol fire from NOAA-AVHRR data (1991). From left to right and from top to bottom; 28 July, 29 July, 30 July, 31 July, 1 August, 3 August (night), 3 August (day), 4 August (night) and 4 August (day). All the images but the 6th and 8th were acquired from the afternoon pass of the satellite (around 2.30 p.m). The other two were acquired at the night overpass (around 2.30 a.m).

are based on detecting the active fire in each of the multi-temporal images covering a single event (Fig. 6.2). Each daily image is converted into a binary mask (fire/not fire). Afterwards, the images are co-registered and superimposed. The pixels detected as fire are labelled according to the date and time of image acquisition. The overlay of the daily series makes it possible to obtain a single file, in which the spatial evolution of the fire can be portrayed (Plate 6.1). This map synthesises the spatial evolution of the fire for the days when images were available. In spite of the coarse spatial resolution of the AVHRR data, the map matched the information collected in the field quite well. Future improvements could include the use of sub-pixel detection techniques to improve the spatial resolution of the results.

6.6
Future systems

There are currently several new satellites and sensors under design and construction that will provide in the next few years further capabilities for the monitoring of fire and related issues. Some of these instruments will include modifications and improvements to existing systems, whilst others will offer completely new sensing capabilities, complementing each other. We will review here the most promising ones for fire detection.

The next generation of geo-synchronous satellites will offer improvements in the ability to sense biomass burning activities and to track associated emissions. The first of the METEOSAT Second Generation (MSG) satellites will improve capabilities for real-time fire monitoring over Europe and Africa. The new imaging sensor on board this platform, the Spinning Enhanced Visible and Infra Red Imager (SEVIRI), consists of a set of 12 spectral channels, including channels identical to AVHRR's, providing full disk coverage of Europe, Africa and the Atlantic Ocean every 15 minutes. Of particular interest will be the spectral channels centred at 1.6 and 3.8 μm. These channels will be available at the resolution of 3 km at sub-satellite point. The sensor will also include a broadband High Resolution Visible channel with a spatial resolution of 1 km. The first satellite MSG-1, is slated for launch in late 2000, and will then start to be operationally available from early 2001. (See the Internet address of Eumetsat: http://www.eumetsat.de for more information).

The Moderate Resolution Imaging Spectro-radiometer (MODIS) on board the NASA's Earth Observing System (EOS) has been designed to include specific characteristics for fire detection. It will fly in a polar orbit and will provide global coverage between 1 and 2 days. This new instrument will have 36 spectral channels between 0.4 and 14.3 μm with spatial resolutions from 250-1000 metres. Of particular interest to fire monitoring are (i) for day time monitoring, two high gain channels located at 3.9 and 11 μm with saturation levels at 500 K and 400 K respectively (1 km resolution), (ii) for night time the 0.86 μm (250 m resolution), 1.65 and 2.15 μm (500 m resolution) channels. MODIS will provide fire products including information on the accurate location of a fire, its emitted energy, the flaming and smouldering ratio and an estimation of area burned. EOSDIS will distribute daily these 1 km full resolution fire products, as well as spatially and temporally summarised information for scientists requiring data on fire distribution, timing and characteristics for inclusion in modelling studies. Besides this, the MODIS will offer improved bands for cloud detection and atmospheric correction. The EOS-AM platform is planned for launching in mid-1999, and will be followed by the PM platform allowing four observations daily (See the Internet address of NASA: http://modarch.gsfc.nasa.gov/ for more information).

In the continuation of ATSR, which demonstrated capabilities in fire monitoring (Perrin and Millington, 1997), the Advanced Along Track Scanning Radiometer (AATSR) will be carried on the European Space Agency (ESA) platform ENVISAT-1, to be launched in 1999. AATSR will provide two channels useful to fire monitoring located at 1.6 and 3.75 μm. It will continue to offer daily global coverage at 1×1 km resolution at nadir (500 km swath). See the Internet address of ESA (http://envisat.estec.esa.nl/instruments/aatsr/ for more information).

In recent years, the number of specifically dedicated remote sensing programmes has increased. In the field of fire detection one of the most promising is the so-called FUEGO programme, founded by the European Commission. It has been defined with the specific objective of defining and developing a constellation of 9 to 12 satellites to provide operationally:

- Early forest fire detection compatible with fire fighting requirements with direct signalling to the forces.

- Forest fire high resolution monitoring providing a synoptic fire line direct to the fire chief.
- Additional value information like short-term risk prediction, hot spot identification and burned surface assessment.

The system will cover all the fire prone areas world-wide and it is being tailored to cover especially Mediterranean Europe, the USA, Australia and South America. The FUEGO system will provide high accuracy fire locating characteristics and wide coverage fire monitoring capability. A low Earth orbit (LEO) constellation will enable the onboard infrared instruments to scan the fire risk areas for outbreaks of wildfires. The orbit has an inclination of 47.5° in order to optimise the revisit time for European latitudes, which are similar to other fire prone areas in the world (California, Southeast Australia, Chile, Argentina, etc). This requires a width of around 2500 km, centred at the sub-satellite point and perpendicular to the velocity direction. FUEGO satellites will carry on board infrared detectors in both the 3-5 range and 8-10 μm range. Auxiliary channels in the visible and near infrared parts of the spectrum will also be included to improve false alarm filtering and generate complementary information. The satellites and the subscription terminals should be available around 2002-2003. (See de Internet address of the Fuego program http://ctv.es/insa/fuego/home.htm for more information).

Within the same context of sensors designed specifically for fire research, the *Deutsches Zentrum für Lüft- und Raumfahrt* (DLR) is preparing a small satellite mission, which will launch a Bi-spectral IR Detection instrument (BIRD). The satellite instrument will comprise two infrared sensors at $\lambda_1 = 3.7$ μm and $\lambda_2 = 8.8$ μm and a VIS/NIR wide-angle stereo scanner. The mission objective of BIRD is to be a precursor for European space-borne high-temperature-event-recognition systems such as FOCUS, FUEGO and Forest Fire Earth Watch. The launch of BIRD is planned for 1999.

FOCUS is an Intelligent Infrared Sensor System, also proposed to the European Space Agency (ESA) by DLR, as a prototype experiment for a high-temperature disaster recognition satellite system. Still at an early stage, the FOCUS mission is foreseen as experimental research that will pave the way for future operational fire detection missions.

6.7
Conclusions

This chapter explained how middle infrared data can be successfully used to detect fires from space. Of existing systems offering information in that part of the spectrum, the NOAA-AVHRR sensors have been shown to provide the most appropriate information for fire fighting decision making in the Mediterranean areas. They present the best trade-off between spatial and temporal resolution, while allowing direct reception and near-real time dissemination of fire locations. However, specialised algorithms are required to select active fires from the data because AVHRR was designed for meteorological purposes. We have reviewed such algorithms, of which those based on the contextual approach are the present state-of-the-art.

Further development in such algorithms may benefit from improved knowledge of:

- The effect of temperature in-homogeneity within the fire on the pixel measurement.
- The effect of fire size, temperature and backgrounds on the pixel measurement.
- Exhaustive ground data for validation purposes.

While AVHRR data will improve operational availability of near real-time fire location information to fire fighters in the field in Mediterranean areas, substantial benefit will probably only come with the arrival of the next generation of sensors. Specially designed for fire issues, they will offer increased spectral information, offer higher saturation levels for the MIR channels and improve temporal and spatial resolutions.

7 Spectral characterisation and discrimination of burnt areas

José M.C. Pereira, Ana C.L. Sá, Adélia M.O. Sousa, João M.N. Silva,
Teresa N. Santos, and João M.B. Carreiras
Departamento de Engenharia Florestal, Instituto Superior de Agronomia (Portugal).

Abstract. Spectral properties of recent burns are characterised, in the visible, near infrared, mid-infrared, thermal infrared, and microwave spectral domains. Fire-induced reflectance changes are also compared for various ecosystems and biomes, and discussed in terms of the ecological effects of phytomass combustion. The spectral signatures of combustion products and of burnt areas are compared with those of various plant material and land cover types, in order to graphically represent relevant aspects of burnt area spectral discrimination. A series of colour composite images, based on Landsat Thematic Mapper data is used to illustrate the appearance of burnt surfaces in various tri-spectral spaces, and in contrast with healthy forests, agricultural fields, and urban areas. The temporal evolution of the spectral properties of burns is also demonstrated, with a five-year time series of Thematic Mapper images of two conifer forest burns in central Portugal. Finally, a series of conclusions is proposed, concerning the distinctive spectral properties of burnt surfaces, and implications for discrimination and mapping of such areas.

7.1 Introduction

Vegetation fires are common in tropical, temperate, and boreal biomes. In the tropics fire is often used as a land management tool, employed in shifting agriculture, hunting practices, to prevent the invasion of grasslands by shrubs, and in the conversion of primary forest to other uses. In temperate regions fire is also extensively used for slash burning, but is most often considered a hazard when it burns in ecosystems that are prized for their economic and ecological value. In both tropical and temperate biomes, a great majority of fires, whether desired or not, occur as a result of human activities, but in boreal regions natural causes (i.e. lightning) still account for a significant proportion of the total area burnt, and fire is a very important ecological factor influencing plant community dynamics.

These various kinds of fire activity generate a range of economic, ecological, atmospheric and climatic impacts, with magnitudes that are strongly dependent on the areal extent of the burns. Detailed and current information concerning the location and extent of the burnt areas is important for assessing economic losses and ecological effects, monitoring land use and land cover changes, and modelling atmospheric and climatic impacts of biomass burning. Given the very broad spatial extent and often limited accessibility of the areas affected by fire, satellite remote sensing is an essential technology for gathering the required information in a timely and methodologically consistent manner.

However, the spectral properties of burnt surfaces are prone to confusion with various land cover types, affecting the accuracy of burnt area estimates derived from remotely sensed data. The development of methodologies capable of producing more accurate burnt area estimates from remotely sensed data is, therefore, an active topic of research at geographic scales ranging from local to global (Justice et al., 1993).

7.2
Spectral properties of burnt areas

Spectral characterisation of the post-fire signals was considered by Chuvieco and Congalton (1988) as the starting point for research on remote sensing of burnt areas. An essential aspect of the problem is the recognition that there are two quite different post-fire signals (Robinson 1991): the formation and deposition of charcoal, or surface charring, and the alteration of vegetation structure and abundance, commonly designated by fire scar. The first type of signal is a quite unique consequence of vegetation combustion, but has relatively short duration and tends to be strongly attenuated by wind and rainfall within a few weeks or months after the fire. The second signal is more stable (although its persistence may vary from 2-3 weeks in tropical grasslands, to several years in boreal forest ecosystems), but is less significant to discriminate fire effects, since partial or complete removal of plant canopies may also be due to other factors such as cutting, grazing, wind throw, water stress, or the action of insects and pathogens.

This fundamental distinction seldom is explicitly recognised, leading to apparent inconsistencies in the literature concerning the spectral properties of burnt areas. Most authors, however, indicate the length of time between fire occurrence and spectral data acquisition, from which the type of signal (char or scar) may be inferred approximately, by taking into account biome-specific differences in post-fire spectral dynamics.

It is also important to make a clear distinction between ash and char or, using the terminology of Chandler et al. (1983), "white ash" and "black ash". Ash is a light coloured, predominantly mineral residue, produced by complete combustion of plant materials in the presence of unrestricted oxygen supply, and as a result of high fire intensity (Cope and Chaloner 1985; Riggan et al. 1994). Charred fuels are essentially composed of graphitic carbon, indicate inefficient combustion of biomass under more restricted oxygen supply conditions (Cope and Chaloner 1985), and are the typical product of less severe wildfire behaviour (Chandler et al. 1983; Ambrosia and Brass 1988). Unfortunately, both of these solid products of biomass burning are often designated by "ash", thus confusing interpretation of the spectral properties of burnt surfaces. In the following sections we review the main spectral characteristics of burns, from the visible to the microwave range, using data from our own work, and also from a non-exhaustive but illustrative literature review.

7.2.1
Visible (0.4 - 0.7 μm)

This spectral domain is one of the better known and more extensively used for the purpose of burnt area mapping. Results available in the literature concern both field radiometry work, and analyses of satellite imagery from various sensors, with emphasis on the Landsat Thematic Mapper (TM), and the National Oceanic and Atmospheric Administration's Advanced Very High Resolution Radiometer (NOAA/AVHRR). A wide range of biomes and ecosystems is also covered, including boreal forests, temperate forests and shrublands, and tropical forests, woodlands and savannas. Table 7.1 presents a summary of literature results on the spectral properties of burnt surfaces in the visible range.

Recently burnt surfaces tend to be relatively dark in the visible range, and their reflectance exceeds 0.1 in only three of the studies analysed. The very high reflectance values reported for one to three year old fire scars by Siljeström and Moreno (1995) is possibly due to the contribution of a bright sandy soil background, especially for the older burns. An interesting fact is the typical fire-induced increase in visible reflectance reported by most studies. This is possibly due to the fact that green vegetation in forest and shrublands is considerably dark in this spectral domain. If even moderately bright soil is exposed due to the fire, the overall surface reflectance will increase. Exceptions are the field radiometry works of Frederiksen et al. (1990) and Langaas and Kane (1991). These authors measured the reflectance of very recent fire scars in tropical savannas during the dry season, and the observed darkening of the surface, in these cases, is relative to dry grass, which has much higher visible reflectance than green forests and shrublands. Jakubauskas et al. (1990) also observed lower post-fire reflectance in a mixed conifer-hardwood forest, but only in the most severely burnt areas. Several authors indicate that the visible spectral range is not very effective for discriminating burns (Tanaka et al. 1983; Ponzoni et al. 1986; López and Caselles 1991; Pereira and Setzer 1993; Koutsias and Karteris 1996; Silva 1996; Pereira 1999). A few reasons for this are suggested by Pereira (1999), namely: (1) Like recent burns, several common land cover types, namely water bodies, wetlands, dense conifer forests, and many soil types are quite dark in the visible. These similarities reduce the possibility of using the visible range to discriminate burns; (2) The dynamic range available with Earth observation satellites for discriminating between these different types of surfaces, all of which are dark in the visible, is narrow; (3) path radiance, an important component of the atmospheric effect, predominates in the visible range especially over dark surfaces, and causes a loss of contrast between different land cover types.

Table 7.1. Spectral properties of burnt surfaces in the visible range

Reference	Vegetation type	Data type	Spectral region	(ρ)	Spectral change	Comments
Fuller and Rouse 1979	Boreal forest	Surface ρ	0.4 μm 0.7 μm	0.025 0.05	Slight increase	Field radiometry. 0 to 1 year old burns
Tanaka et al. 1983	Pine branches	Surface ρ	0.4 μm- 0.7 μm	0.09 0.14	0.05-0.08 increase	Field radiometry

Ponzoni et al. 1986	Brazilian cerrado	DN	TM3	-	6-8 DN increase	-
Chuvieco and Congalton 1988	Pine forests and shrubland	Radiance	TM1-TM3	-	Increase	Fire scars a few days old
Frederiksen et al. 1990	W. African savanna	Surface ρ	red	0.054	0.05-0.067 decrease	Field radiometry
Jakubauskas et al. 1990	Mixed conifer--hardwood forest	DN	MSS4[4] MSS5[5]	-	decrease decrease (severe) increase (moderate and light)	Burn classified into severe, moderate, and light
Jones et al. 1991	Spruce, birch and pine charcoal	Surface ρ	0.546 μm	0.00-0.06	-	Laboratory spectroscopy. Charcoal ρ as a function of temperature
Langaas and Kane 1991	W. African savanna	Surface ρ	red	0.05	0.035 decrease	Field radiometry
López and Caselles 1991	Evergreen oak and pine forest, shrubs	TOA ρ	TM1[1], TM2[2] TM3[3]	0.10 0.12	0.04 increase	-
Pereira and Setzer 1993	Tropical forest and pastures	DN	TM1-TM3	-	Increase	In TM2 and TM3 burns are darker than pastures
Caetano et al. 1994	Pine forest and shrublands	DN	TM1-TM3	-	Increase	10 days and 4 month old burns
Siljeström and Moreno 1995	Xerophytic shrubs	TOA ρ	TM3	0.35	0.15 increase	1 to 3 year old scars.
Hlavka et al. 1996	Tropical savanna	DN	AVHR R1[5]	-	Variable	Burns darker than bare soil, similar to vegetation
Koutsias and Karteris 1996	Pine forest and shrublands	DN	TM1-TM3	-	Increase	One week old burn
Silva 1996	Temperate forests and shrublands	TOA ρ	TM1 TM2 TM3	0.095 0.073 0.066	0.01-0.02 increase	3 to 6 month old scars
Pereira 1999	Temperate forests and shrublands	TOA ρ	AVHR R1	0.08	0.02 increase	Less than 4 month old scars

(ρ) = Reflectance ; [1] 0.49 μm; [2] 0.56 μm; [3] 0.66 μm; [4] 0.525 μm; [5] 0.63 μm.

7.2.2
Near-infrared (0.7 - 1.3 μm).

The near-infrared (NIR) spectral region has also been extensively used to analyse burnt surfaces, and most of the studies mentioned in the previous section also report results for the NIR. A summary of results is given in Table 7.2.

Table 7.2. Spectral properties of burnt surfaces in the near-infrared range

Reference	Vegetation type	Data type	Spectral region	(ρ)	Spectral change	Comments
Fuller and Rouse 1979	Boreal forest	surface ρ	1.3 μm	0.15	decrease	0 to 1 year old burns
				0.40	Increase	25 year old scar
Tanaka et al. 1983	Pine branches	surface ρ	1.1 μm	0.17-0.22	0.05-0.10 decrease	-
Ponzoni et al. 1986	Tropical savanna	DN	TM4	-	decrease	Only water was darker than burns
Chuvieco and Congalton 1988	Pine forests and shrublands	radiance	TM4	-	decrease	Fire scars a few days old
Frederiksen et al. 1990	W. African savanna	surface ρ	NIR	0.059	0.13 decrease	1 day old burn
Jakubauskas et al. 1990	Mixed conifer--hardwood forest	DN	TM4	-	decrease in all 3 burn severity classes	Darkening of surface in TM4 was used to delineate fire severity classes
Langaas and Kane 1991	W. African savanna	surface ρ	NIR	0.065	0.155 decrease	Same day as late dry season fire
				0.075	0.115 decrease	1 day old early dry season fire
López and Caselles 1991	Evergreen oak and pine forest, shrubs	TOA ρ	TM4[1]	0.12	0.06 decrease	TM4 was the only channel where post-fire ρ decreased
Pereira and Setzer 1993	Tropical forest and pastures	DN	TM4	-	decrease	Burns remain darkest surfaces during 4 years
Caetano et al. 1994	Pine forest and shrublands	DN	TM4	-	decrease	Only shade endmember was darker than burn endmember
Siljeström and Moreno 1995	Sand dune xerophytic shrubs	TOA ρ	TM4	0.42	increase	1 to 3 year old scars
Hlavka et al. 1996	Tropical savanna	DN	AVHRR2[2]		Decrease	Burns darker than bare soil

						and vegetation, similar to water
Koutsias and Karteris 1996	Pine forest and shrub- lands	DN	TM4	-	decrease	One week old burn
Silva 1996	Temperate forests and shrublands	TOA ρ	TM4	0.067	decrease	Only water was darker than burn
Pereira 1999	Temperate forests and shrublands	TOA ρ	AVHRR2	0.08	0.02 increase	Less than 4 month old scars

[1] 0.83 μm; [2] 0.91 μm.

The NIR is the spectral region where the signal of recent fire scars is strongest, and is generally considered as the best spectral region for burnt area detection and mapping (Hall et al. 1980, Tanaka et al. 1983, Richards 1984, Ponzoni et al. 1986, Frederiksen et al. 1990, Langaas and Kane 1991, López and Caselles 1991, Pereira and Setzer 1993, Caetano et al. 1994, Koutsias and Karteris, 1996, Silva, 1996, Sousa 1998, Pereira 1999), especially when pre-fire fuel loadings are high and combustion produces large amounts of charcoal that are deposited on the ground. Since green vegetation is very reflective in the NIR, burning typically causes more or less significant decreases in reflectance. The only exceptions to this trend in Table 7.2 are the already mentioned study by Siljeström and Moreno (1995) and, in the case of Fuller and Rouse (1979), for a 25-year-old fire scar. Contribution of bright soil backgrounds is a likely explanation for these cases of older fire scars. The phenological status of the vegetation (i.e. green versus dry) prior to the fire appears to be less significant in the NIR, and therefore does not introduce a distinction between the burning of dry savannas on one side, and the burning of green forests and shrublands, on the other side. Both green and dry vegetation have substantially higher reflectances than recent burns, and therefore darkening of the burnt surfaces in the NIR is practically systematic.

Figure 7.1 shows the visible-NIR spectral reflectances of green and scorched maritime pine (*Pinus pinaster* Ait.) needles, and also of charcoal and ash from a pine forest fire. Charcoal is very dark throughout the entire spectral range, while ash is very bright and displays a monotonical increase in reflectance with wavelength. Scorched needles are brighter than green needles in the blue and red wavelengths, and darker in the green and NIR. A fire of moderate severity that only scorches the tree crowns, will cause an increase in visible reflectance, and a decrease in NIR reflectance, relatively to the healthy forest canopy. Plant material reflectance measurements made at the canopy level are known to produce significantly lower reflectances than those shown in Fig. 7.1 (Williams 1991), such that the visible reflectance of a green pine forest canopy is quite close to that of a burnt surface, while the NIR canopy reflectance remains substantially higher. This is illustrated in Figure 7.2, that shows results from a simulation using a hybrid geometrical-optical / radiative transfer (GORT) model (Ni et al. 1998). The spectral signatures of a green maritime pine forest canopy over a pure charcoal background were derived as a function of stand density, ranging from 500 to 2000 trees/ha. Reflectances in the blue and red wavelengths are almost independent of stand density, showing that a dense pine forest canopy is almost as dark as a charred

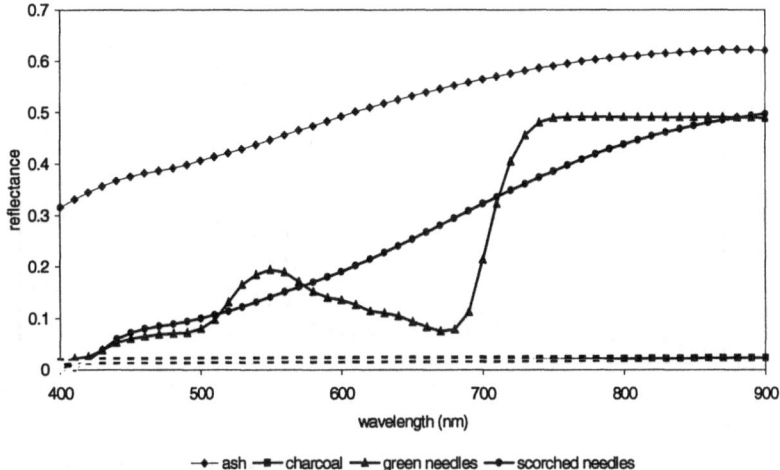

Fig. 7.1. Spectral reflectance signatures of characteristic constituents of fire-affected areas, at a spectral resolution of 10 nm. The data were collected with a portable spectroradiometer under field conditions (charcoal and ash), and in the laboratory using an integrating sphere (needles).

surface at these wavelengths. On the contrary, NIR canopy reflectance increases markedly with stand density

7.2.3
Mid-infrared (1.3 – 8.0 μm)

Observations of the Earth surface in the mid-infrared (MIR) only became widely available after the launch of the Landsat-4, carrying the Thematic Mapper sensor with two channels in this spectral range, TM5 (1.6 μm) and TM7 (2.1 μm). Field radiometric or spectroradiometric data concerning burnt areas in the MIR appear to be unavailable in the literature. Most of the work done on remote sensing of burnt surfaces in the MIR relied on the Landsat TM, with some exceptions such as the studies by Eva et al. (1995) with the ATSR-1, and Pereira (1999) with the AVHRR (Table 7.3).

MIR spectral changes induced by fire are similar to those in the visible range, since burnt areas are typically more reflective than green vegetation, but darker than the predominantly senesced vegetation of tropical savannas during the dry season. Pereira and Setzer (1993) show no exception, because the fire scars were found to be brighter than primary tropical forest, and darker than secondary forest and grasslands.

The differences between temperate forest and shrublands, and tropical savannas regarding the MIR spectral changes caused by burning are likely to result from the phenological factors already mentioned in the case of the visible range. However, in the visible range, the relevant physiological change inducing high pre-fire vegetation reflectance was the loss of chlorophyll, while in the MIR this is a

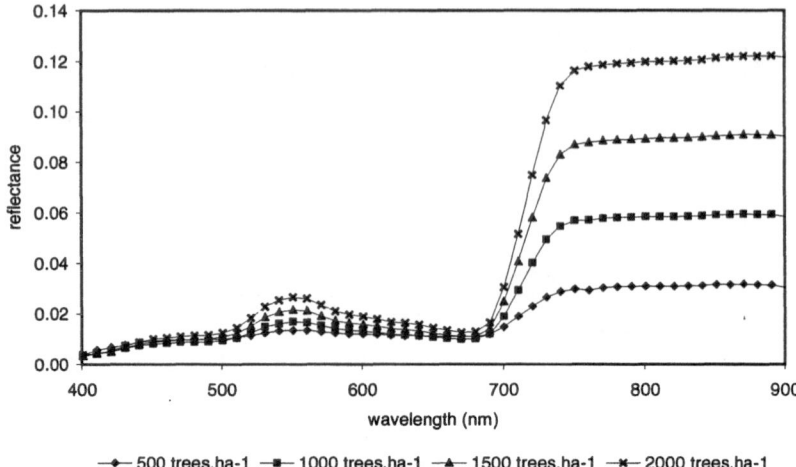

Fig. 7.2. Simulated spectral reflectance signatures of an understory fire that produces a pure charcoal background, but leaves an intact tree canopy. A leaf area index of 3 was assumed for each individual tree, and signatures are simulated as a function of stand density. The GORT model (Ni et al. 1998) was used in the simulation.

consequence of loss of water by the plant tissues. The mid-infrared spectral region has been more recently identified by a few authors, mostly working on temperate ecosystems, as promising for detecting burns, since the fire-induced increase in brightness is larger than in the visible range (López and Caselles 1991; Koutsias and Karteris 1996; Silva 1996; Pereira 1999). Pereira (1999) found that the increase in reflectance over burnt surfaces is higher in the MIR than in the visible, and thus considered the NIR/MIR bispectral space more appropriate for burnt area discrimination and mapping than the classical visible/NIR space used in remote sensing of vegetation. The Landsat TM MIR channels were identified by some authors as those where burned surfaces displayed a larger scatter of brightness values (Ponzoni et al. 1986; Pereira and Setzer 1993; Koutsias and Karteris 1996; Silva 1996), and Pereira (1999) presented similar results for the reflective component of the AVHRR channel 3. This suggests the possibility of using the MIR for characterising the internal variability of burned areas, which may be correlated with the severity of fire effects, and the spatial pattern of fire intensity (Justice et al. 1993). Also, atmospheric scattering, including for smoke aerosols is very small in this wavelength range, and therefore does not reduce spectral contrast at the surface.

Figure 7.3 shows top-of-atmosphere (TOA) spectral reflectance of several different types of surfaces, using data from the Landsat 5 TM. The spectral signatures displayed represent mean values of several training areas extracted from over 20 Landsat TM scenes in Portugal. Two types of burnt surfaces are shown, one corresponding to a fire in a shrubland, and another to a fire in a conifer forest. The forest fire scar has slightly higher reflectance in the visible and NIR, and slightly lower reflectance in the MIR, but the two signatures are quite similar. MIR reflectance of the burnt surfaces is lower than that of unvegetated surfaces (bare soil, rock outcrop, and urban area), but higher than the reflectance of vegetation and water. Contrary to vegetation, recent burns display a small "NIR edge", a characteristic rise in reflectance from the NIR to the MIR. This feature is also present in the signatures of other unvegetated surfaces, but burns tend to have lower reflectance than these land cover types over the entire spectrum of the TM.

Table 7.3. Spectral properties of burnt surfaces in the mid-infrared range

Reference	Vegetation type	Data type	Spectral region	(ρ)	Spectral change	Comments
Ponzoni et al. 1986	Tropical savanna	DN	TM5 TM7	-	decrease	Burns are darkest surface in TM5 Only water is darker in TM7
Chuvieco and Congalton 1988	Pine forests and shrublands	radiance	TM5 TM7	-	increase	Fire scars a few days old
López and Caselles 1991	Evergreen oak and pine forest, shrubs	TOA ρ	TM5[1] TM7[2]	0.22 0.18	0.05 increase 0.10 increase	TM7 reflectance drops to pre-fire levels only 6 years after fire
Pereira and Setzer 1993	Tropical forest and grasslands	DN	TM5 TM7	-	variable	Burns were brighter than primary forest, but darker than secondary forest and grasslands
Caetano et al. 1994	Pine forest and shrublands	DN	TM5 TM7	-	increase	Burn was brighter than vegetation endmember, but darker than soil endmember
Eva et al. 1995	Tropical savanna and forest	DN	ATSR-1 SWIR[1]		decrease	Time series analysis of dry grass fires
Siljeström and Moreno 1995	Xerophytic shrubs	TOA ρ	TM5 TM7	0.90 0.55	0.50 increase 0.35 increase	1 to 3 year old scars
Koutsias and Karteris 1996	Pine forest and shrublands	DN	TM5 TM7		Stable increase	One week old burn
Silva 1996	Temperate forests and shrublands	TOA ρ	TM5 TM7	0.153 0.141	increase	Only bare soil was brighter Burn was brightest surface
Pereira 1999	Temperate forests and shrublands	TOA ρ	AVHRR 3[3]	0.1	0.04 increase	Used reflective component of AVHRR channel 3

[1] 1.6 μm; [2] 2.1 μm; [3] 3.75 μm

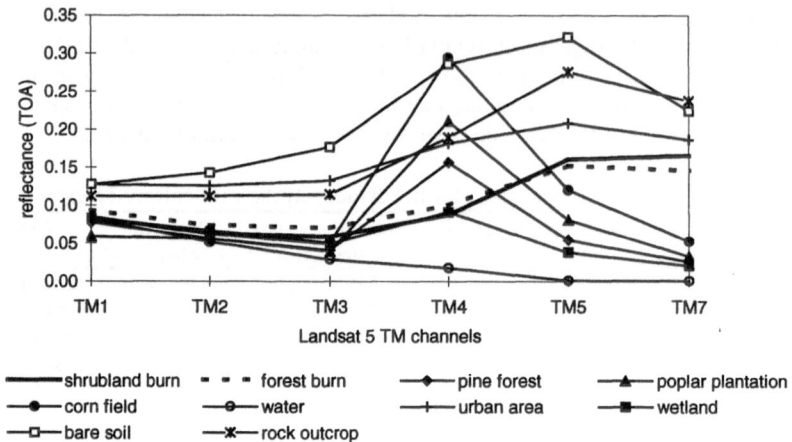

Fig. 7.3. Spectral signatures of diverse land cover types, extracted from Landsat TM imagery. The data were not corrected for atmospheric effects.

Figure 7.3 also shows that both types of burn have slightly higher reflectance than green vegetation in the red channel (TM3). In the green channel (TM2) the reflectances of vegetation and burns overlap one another, and in the blue channel (TM1) the poplar plantation and the corn field are darker than both burns.

7.2.4
Thermal infrared (8.0 – 14.0 μm)

In this section we do not consider the thermal signal generated during the fire, by flaming or smouldering combustion of phytomass, i.e. the active fire signal. Our discussion focuses on the post-fire surface heating, considered by Whelan (1995) as one of the most relevant changes induced by fire in the physical environment. Surface heating is caused by a combination of factors, such as the removal of vegetation shading and of the insulative effect of litter, lowered albedo of the soil surface, and altered soil water relations (Whelan 1995).

A diversity of sensors have been used to analyse burnt surfaces in the thermal infrared, namely the Landsat TM, the AVHRR, and the ATSR (Table 7.4), and all authors report the expected increase in temperature over burnt surfaces, relative to the pre-fire situation. However, some authors consider the temperature differences too small to allow the reliable detection of burns (Pereira and Setzer 1993), while others mention that the differential thermal signal vanishes rapidly as soon as the vegetative cover starts to regenerate (López and Caselles 1991). Nevertheless, Eva et al. (1995), working with a time series of daily ATSR-1 data in the tropical savannas of Central Africa, found the temperature increase, in conjunction with a decrease in MIR reflectance, to be a reliable detector of burnt surfaces.

The usefulness of thermal data for the detection of burnt surfaces appears to be variable, depending on the degree of contrast with surrounding land cover types, the time since the fire (i.e. scar age), and the meteorological conditions during the period between fire occurrence and data acquisition. However, these caveats also apply to the analysis of data from the solar reflective spectral domain.

Table 7.4. Spectral properties of burnt surfaces in the thermal infrared range

Reference	Vegetation type	Data type	Spectral region	Thermal change	Comments
López and Caselles 1991	Evergreen oak and pine forest, shrubs	Brightness temperature	TM 6[1]	5-6°C increase	One month old burns
Pereira and Setzer 1993	Tropical forest and pastures	DN	TM 6	increase	
Cahoon et al. 1994	Boreal forest	Brightness temperature	AVHRR 4[2]	increase	Global Area Coverage data (4 km spatial resolution)
Eva et al. 1995	Tropical savanna And forest	Brightness temperature	ATSR-1 TIR[3]	10-15 °C increase	Time series analysis of dry grass fire scars
Hlavka et al. 1996	Tropical savanna	DN	AVHRR 3[4]	increase	Burns similar to bare soil, brighter than vegetation

[1] 11.5 µm; [2] 11 µm; [3] 10.85 µm; [4] 3.75 µm.

7.2.5
Microwave (> 1 mm)

Most of the work performed so far on the remote sensing of burnt areas in the microwave spectral range has analysed conifer forest ecosystems, mostly in boreal regions. An exception is the study by Malingreau et al. (1995) of a hypothetical savanna fire in central Africa. Spaceborne and airborne synthetic aperture radar (SAR) have been the instruments of choice, C-band the most widely used spectral channel. Several authors have used data from the ERS-1 SAR instrument, which relies on vertical-vertical (VV) polarisation for the C-band, but studies that relied on airborne instruments had access to alternative forms of signal polarisation (Table 7.5).

Apparently contradictory results have been reported by different authors. Kneppeck and Ahern (1989) found a decrease in backscatter intensity over a conifer forest burn, which was more noticeable with HV than with VV polarisation. Kasischke et al. (1992) and Landry et al. (1995) also found a decrease in return intensity over a three year old burn, and a one year old burn, respectively. But Werle et al. (1991) and Landry et al. (1995) reported that the backscatter changes over very recent burns (i.e. about one week old) were too small to allow for detection of the fire scars. On the other hand, Kasischke et al. (1992,1994), Bourgeau-Chavez et al. (1994, 1995), French et al. (1994), and Malingreau et al. (1995) all observed fire-induced increases in backscatter intensity.

Table 7.5. Spectral properties of burnt surfaces in the microwave range

Reference	Vegetation type	Spectral region	Polarisation	Back-scatter change	Comments
Kneppeck and Ahern 1989	Pine and spruce forest	SAR C-band[1]	HV and VV	decrease	Larger change with HV than VV
Werle et al. 1991	Boreal forest	SAR C- and X-band[2]	HH and HV	inconclusive	Burn less than one week old
Kasischke et al. 1992	Boreal forest	SAR C-band	VV	variable	Backscatter increase over recent burns, decrease over 3-year old burns
Bourgeau-Chavez et al. 1994	Boreal forest	SAR C-band	VV	increase	Permafrost melting increases soil moisture
French et al. 1994	Boreal forest	SAR C-band	VV	increase	Permafrost melting increases soil moisture
Kasischke et al. 1994	Boreal forest	SAR C-band	VV	increase	Spatial heterogeneity of signal within burn
Bourgeau-Chavez et al. 1995	Boreal forest	SAR C-band	VV	increase	Permafrost melting increases soil moisture
Landry et al. 1995	Conifer forest	SAR C-band	HH	inconclusive decrease	Very recent burn One year old burn
Malingreau et al. 1995	Tropical shrubby savanna	SAR C-band	VV	increase	No ground checking. Burn is hypothesis

[1]3.75 – 7.5 cm; [2]2.4 – 3.75 cm.

The explanations proposed for the backscatter changes detected invoke alterations in vegetation structure, and in vegetation and soil moisture content. The ecological impacts of fire in the boreal forests of Alaska were thoroughly analysed, with microwave imagery and in the field, by Kasischke et al (1992, 1994),

Bourgeau-Chavez et al. (1994, 1995), and French et al. (1994). These authors found that the consistent increase in backscatter intensity over burn scars was due to an increase in soil moisture caused by permafrost melting, as a consequence of increased solar irradiance and absorbance at the ground level, due to lowered surface albedo.

7.2.6
An overview of the characteristics of burnt surfaces using Landsat 5 TM imagery

7.2.6.1
Spectral properties and colour composites

Plate 7.1 displays a series of Landsat 5 TM red-green-blue (RGB) colour composites, showing burns that occurred during the summer of 1995, in a diverse landscape of central Portugal. The satellite image dates from September 24, a few weeks after the fires. The region shown covers an area of approximately 30km by 30 km, where the urban area identified with the number 2 is the city of Coimbra. The colour composites were chosen to show different views of this landscape in alternative three dimensional spectral spaces. The first image (Plate 7.1a) is a "true colour" view, with TM channels 3, 2, and 1 in the red, green, and blue colour planes (RGB-321), a representation that uses spectral data only from the visible range. The burns appear as brown patches (1), clearly darker than urban areas (2) and agricultural vegetation (3), but brighter than forested areas (4), namely the region of rugged terrain in the lower left corner of this subscene. This is consistent with the findings reported in Section 2.1 regarding fire-induced spectral changes in the visible wavelengths.

Plate 7.1b is an RGB-432 composite, with a colour coding similar to that of a false colour infrared aerial photography, thus combining data from the visible (TM2 and TM3), and near-infrared (TM4). Here, the burns appear as dark grey patches, while the vegetated areas are intensely red. Urban areas and bare soils appear in bluish grey tones. The much stronger contrast between burns and vegetated areas, in comparison to plate 7.1a is due essentially to the contribution of the near-infrared channel, TM4. Since vegetation is quite reflective in this spectral region, its removal or drastic reduction by combustion causes a marked decrease in reflectance. The red and NIR spectral data contained in this colour composite define the bispectral space typically used for vegetation studies, and namely for the definition of most vegetation indices.

Plate 7.1c displays an RGB-743 colour composite, thus containing data from the three major spectral regions covered by the TM: visible (TM3), near-infrared (TM4), and mid-infrared (TM7). Burnt areas stand out very clearly as dark red patches, contrasting with the green shades of vegetated areas, and the white to pink tones of urban areas and bare soils. Burnt areas appear dark red because of their relatively higher reflectance in the MIR, relatively to the visible and NIR. Vegetated areas are predominantly green because they are highly reflective only in the NIR (TM4), since plant pigments absorb strongly in the visible (TM3), and water contained in plant tissues absorbs MIR (TM7) radiation. Recent work by several authors has shown that spectral indices defined in the NIR/MIR bispectral space (a subset of this colour composite) are more efective for detecting burnt surfaces than the traditional red/NIR bispectral space (López and Caselles 1991, Silva 1996, Sousa 1998, Pereira 1999).

Finally, Plate 7.1d shows an RGB-647 composite, constructed with thermal infrared data (TM6), NIR data (TM4) and MIR data (TM7). This composite excludes any visible channel data. The resulting image is quite similar to the previous one, but burnt surfaces appear now as patches of a brighter tone of red. This is due to the fact that these surfaces are hot and dry, and therefore both TM6 and TM7 contribute significantly to their signal. Use of the thermal channel also appears to introduce additional spectral discriminant ability for distinctive vegetation types, as seen in the lower left hand corner of the subscene.

7.2.6.2
Fire-induced spectral changes and vegetation recovery

Changes in the spectral properties of vegetated areas caused by fire were analysed in the previous sections, for the main spectral regions commonly used in satellite-based Earth observation. Plate 7.2 illustrates these concepts with a time series of six Landsat 5 TM images (RGB-743) of a site in central Portugal, located about 35 km west of the region shown in Plate 7.1. This area covers approximately 43 km by 20 km, containing a coastal mountain, Serra da Boa Viagem, and a sand dune pine forest. Two fires took place in this area during the summer of 1993, one that burned most of the sand dune pine forest (identified as 1 in Plate 7.2b), and a smaller one, that affected Serra da Boa Viagem (identified as 2 in Plate 7.2b).

The spectral changes caused by the fires were quite drastic, as can be seen in Plate 7.2a, dated from November 1992 and Plate 7.2b, from October 1993. The burnt areas appear in red colour, brighter in site 1 and darker in site 2. This difference is probably due to the brighter soil colour in the sand dune, and to the effect of topographic shading, in the mountain site. In Plate 7.2c (December 1994), a marked difference in the rates of vegetation recovery between the two sites is already evident. The burn scar in site 2 is barely visible, partly due to deep topographic shading, while in site 1 the scar still displays a very strong contrast with the surrounding landscape. The fire scar on the sand dune evolves into progressively lighter tones of pink, in Plates 7.2c, 7.2d (September 1995), and 7.2e (March 1996), possibly due to the gradual removal of combustion residues, and increasing exposure of the bright sandy soil background. It is only five years after the fire (Plate 7.2f, October 1997) that vegetation recovery becomes apparent in site 1. Differences in primary productivity between the two sites are also probably related to the very distinct rates of post-fire vegetation recovery. A thorough discussion of factors influencing post-fire successional patterns in Mediterranean-type ecosystems, and their assessment with Landsat TM data is available in Viedma et al. (1997). Detailed comparison of colour patterns between the different dates must be done with caution, taking into account that the images were not calibrated, neither in absolute nor in relative terms. However, equivalent histogram stretching parameters were used, in order to allow for a visual, qualitative comparison of the various dates. Figure 7.4 displays the temporal trajectory of the median value of a data sample from fire 1 (marked with a yellow square in Plate 7.2b), in the spectral space of the colour composites. The change from 92 to 93-94 reveals essentially a loss of the vegetation signal. From 93-94 to 95, there is a migration along the soil line (in TM3/TM4 space), towards a brighter background, corresponding to charcoal removal, and from 95 to 96 there is evidence of some recovery of the vegetation signal. Finally, in 97 the median pixel appears vegetated but more reflective in the NIR (TM4), indicating the presence of a "shallower" canopy typical early stages of post-fire vegetation recovery.

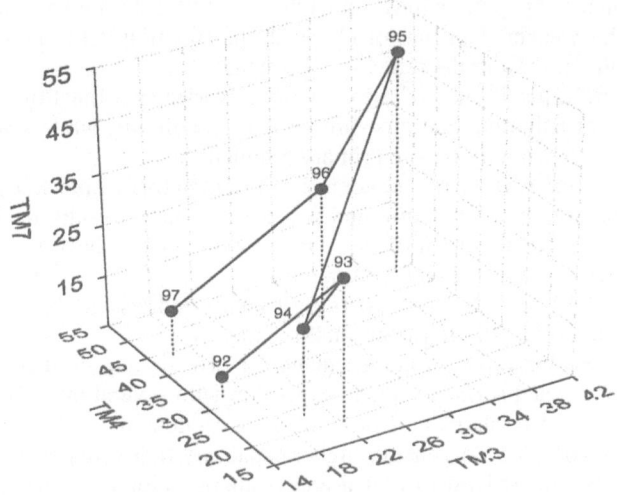

Fig. 7.4. Temporal trajectory of a fire scar data sample in red-NIR-MIR (TM3-TM4-TM7) spectral space, over a 5-year period, from November 1992 (pre-fire) to October 1997.

7.3
Conclusions

Based on our own experience, complemented by a review of the literature on detection and mapping of burnt areas using field radiometry data and satellite imagery, it was possible to draw a series of conclusions. The degree to which these conclusions can be generalized, and the firmness with which they are established is, of course, variable. Scientists working in specific regions and dealing with concrete data sets should take them as tentative, and assess to what extent they are confirmed or denied by their own data and insights. Nevertheless, we propose the following list of conclusions:

1. In the visible spectral range, there appears to be a difference between the short-term post-fire response of boreal and temperate ecosystems, on the one side, and of tropical ecosystems, on the other side. In the former, recent burns tend to be brighter than pre-fire vegetation, while in the latter they generally appear darker.

2. This distinction does not occur in the NIR region, where the brightness of recently burnt areas is systematically lower than that of the pre-existing land cover types.

3. Spectral response to fire in the MIR is similar to that observed in the visible region of the spectrum, but with a larger increase in brightness, in boreal and temperate ecosystems.

4. The thermal signal of burnt areas has not been widely used, although higher temperatures have been observed for recent burns. The signal seems to be weaker than that of charcoal, and possibly less persistent than that originating from the reflective wavelengths.

5. Overall, persistence of the char signal varies as a function of fuel particle size distribution, and post-fire weather conditions, while scar duration depends essentially on site primary productivity.

6. In the microwave range, control of the radiative response by plant canopy structure, and vegetation and soil water content results in diverse, apparently contradictory findings, which in some cases have not yet been thoroughly interpreted and confirmed.

7. The NIR provides the best (i.e. strongest and less equivocal) discriminant ability to identify burns in satellite imagery.

8. The visible range is not adequate for effective spectral discrimination of fire-affected surfaces, due to confusion with several other land cover types that are also relatively dark in this spectral domain.

9. The MIR has, in general, higher capability to identify burns than the visible range. At least in boreal and temperate biomes burns appear brighter, and this spectral region is also much less sensitive to atmospheric disturbances.

10. As a consequence of the last three conclusions, the NIR / MIR bispectral space has a stronger discriminant ability for recently burned surfaces than the classical NIR/visible space.

11. Burnt surfaces reveal the ecological effects of fires, and these affect their levels of greenness, albedo, temperature, and soil moisture. Therefore, improved methodologies for burnt area mapping will require the selection, development and integration of indices sensitive to the spectral characteristics of these surface parameters.

Acknowledgements. We are grateful to several people who collaborated with us in developing the research reported herein, namely: Manuela Batista (Direcção-Geral das Florestas, Lisbon, Portugal), Maria J. Vasconcelos and Mário Caetano (Centro Nacional de Informação Geográfica, Lisbon, Portugal) and Lurdes-Bugalho (Instituto de Meteorologia, Lisbon, Portugal).

8 Regional-scale burnt area mapping in Southern Europe using NOAA-AVHRR 1 km data

José M.C. Pereira, Adélia M.O. Sousa, Ana C.L. Sá
Departamento de Engenharia Florestal, Instituto Superior de Agronomia (Portugal).

M. Pilar Martín and Emilio Chuvieco
Departamento de Geografía, Universidad de Alcalá (Spain).

Abstract. A brief review of studies dealing with burnt area mapping using coarse spatial resolution satellite imagery is presented, followed by an analysis of areas burnt in Iberia during the 1991 and 1994 fire seasons, two of the worst on record in Portugal and Spain, respectively. In order to detect and map burnt areas, new multitemporal image compositing algorithms were developed. Burnt area mapping for the 1991 fire season relied on the Global Environment Monitoring Index (GEMI), albedo, and surface temperature. A rule-based classifier was induced from training data, using the classification and regression trees (CART) algorithm. Fire size estimates compared well to those derived with Landsat TM imagery, but appear unreliable for burns smaller than about 2000 ha. The 1994 fire season data were analysed with a two-phase procedure. First, a new spectral index was specifically designed for burnt area detection. Images of this index were thresholded to detect clearly burnt pixels and to avoid commission errors. Secondly, a distance-based multicriteria analysis technique was applied, combining spectral similarity and spatial contiguity criteria, to map the burns. The method detected over 80% of all large fires, and proved especially effective at mapping burns larger than 1000 ha.

8.1 Introduction

Southern European wildfires affect thousands of square kilometres of forests, shrublands and grasslands every year, causing economic losses, environmental damage and, often, loss of human life. Under the Mediterranean climatic conditions prevailing in the region, the fire season occurs in the summer months. Winter fire regimes are a minority, and can be found mostly in northern Italy and northern Spain. The concentration of a large proportion of the total area burnt in a small number of large fires has been described for the region (Vázquez and Moreno 1995; CE 1996). This agrees with previous findings for southern California and northern Baja California (Minnich 1983; Strauss et al. 1989) and Canada

(Stocks 1991) and is favourable from the standpoint of using low spatial resolution satellite imagery to map burnt areas in southern Europe. It means that detection and mapping of only the larger fires will represent the great majority of the total area burnt every season. Several satellite systems currently available are adequate for this task, such as the NOAA Advanced Very High Resolution Radiometer (AVHRR), the VEGETATION instrument on board SPOT-4, and the Along Track Scanning Radiometer (ATSR), on board ERS-2. The spatial resolution of these instruments is 1 km (VEGETATION and ATSR), and 1.1 km (AVHRR). An important advantage of mapping burnt areas using imagery from one of these systems would be the timely production of burnt area statistics, using a methodologically consistent procedure for the entire region.

To date, most studies of fire using low spatial resolution satellite imagery have relied on the detection of the thermal signal generated by active fires, following the pioneering work of Matson and Dozier (1981) and culminating with recent analyses of daily global fire activity, over a period of almost two years (Dwyer et al. 1998a,b). However, this approach has limitations for burnt area assessment, since short duration and a strong diurnal cycle of the fire thermal signal leads to temporal sampling problems, which can cause severe underestimation of the areas actually affected by fire (Pereira 1990; Pereira et al. 1998a). The feasibility of remotely sensing post-fire signals generated by the effects of fire (vegetation loss, surface charring, heating, and drying), and mapping burnt areas with low spatial resolution imagery, has been demonstrated in recent years for a variety of biomes (Kasischke et al. 1992, 1995; Cahoon et al. 1994; Caetano et al. 1996; Eva and Lambin 1998; Barbosa et al. 1997; Pereira et al. 1998a,b; Pereira, 1999).

8.2
Methods for burnt land mapping

The range of methods used for burnt area mapping with low spatial resolution satellite imagery includes supervised classification (Cahoon et al. 1994; Pereira et al. 1998b), vegetation index image differencing (Kasischke et al. 1992, 1995; Martín and Chuvieco 1995), single-date vegetation index thresholding (Pereira 1999), multitemporal regression analysis (Fernández et al. 1997), fuzzy multiple thresholding (Pereira et al. 1998a; Sousa 1998), spectral unmixing (Caetano et al. 1996), and time-series analysis (Eva et al. 1995; Barbosa et al. 1997). These works are briefly reviewed below.

Kasischke et al. (1993) and Kasischke and French (1995) applied an AVHRR-based multitemporal approach to map forest fires in Alaska, during the 1990 and 1991 fire seasons. Kasischke et al. (1993) created two sets of normalized difference vegetation index (NDVI) maximum value composites (MVC), corresponding to early summer and late summer dates, and differenced the two composites. Burnt areas were identified as those exceeding a given NDVI decrease threshold. This procedure underestimated the area burnt, but generated a large number of single pixel false alarms. However, a less restrictive threshold reduced the area underestimation, and resulted in the detection of 89.5% of all fires larger than 2000 ha.

This methodology was improved by Kasischke and French (1995), who compared an approach based on burnt area estimation using imagery from the same season as the fires, with another approach which also relied on imagery from the following year. Thresholding of the NDVI MVC difference image of the same-year data produced various types of false alarms, requiring elimination of single pixel potential burns, of detections above the treeline, and of single or double pixel detections within 1 km of major rivers. The burnt area was underestimated, in spite of most fires being detected, and use of a less restrictive threshold produced an increase in the number of false alarms. In order to reduce the burnt area underestimation, while minimising the number of false alarms, Kasischke and French (1995) imposed a spatial adjacency constraint on the burnt pixels identified with the less restrictive threshold, and considered as correct only those new detections located within three pixels of the burns identified with the first threshold. The two-year data method was based on the detection of delayed green-up of burnt areas during the first Spring season after the fires. Those pixels with an NDVI value lower than a given threshold below the mean NDVI of unburnt forests, and located within a distance of four pixels from burns detected with the first threshold of the same-year dataset were considered as burns. This method led to 20-30% improvements in burnt area estimation, over the single-year method, at the expense of a delay in the availability of results.

Martín and Chuvieco (1994) employed a method similar to the single-year approach of Kasischke et al. (1993) and Kasischke and French (1995) to analyse wildfires in the Mediterranean coast of Spain, during the Summer of 1991. They also used pre-season/post-season image differencing of AVHRR NDVI imagery, and compared the performance of MVC with analysis of single-date images from the two periods, considering as burns all NDVI decrements larger than 0.2. Image differencing based on two single images was found to match better the available field estimates of fire sizes than MVC differencing, possibly due to residual misregistration between composited images. The area estimation accuracy for the largest fire (15400 ha) using non-composited image differencing was 97.4%. The superior performance of vegetation index image differencing based on single dates, rather than on multitemporal composites was confirmed by Martín and Chuvieco (1995). They also found that simple image differencing performed better than normalised differencing and multitemporal principal components analysis.

Pereira (1999) compared four VI, namely the NDVI, the Vegetation Index 3 (VI3) of Kaufman and Remer (1994), the Global Environment Monitoring Index (GEMI) of Pinty and Verstraete (1992), a modified version of this index, called GEMI3, where the reflective component of channel 3 was used instead of channel 1, and finally channel 2 alone. The new GEMI3, which combines the atmospheric insensitivity of the MIR range with the non-linear design of the GEMI, was found to be the best discriminator for burnt areas, followed by GEMI and VI3, channel 2, and finally the NDVI. This study used a single image dated from the late Summer of 1991, and did not attempt to estimate areas burnt, but concentrated on the comparative analysis of VI performance.

Wildfires in Spain were studied by Fernández et al. (1997), using a time series of 10-day maximum NDVI image composites. They visually identified the approximate location of the central point of burn scars, and defined 100 km by 100 km geographical windows around each centroid. Then performed image differencing and multitemporal regression analyses between pre-fire and post-fire composites. Image segmentation thresholds for burnt area classification were defined at above the mean plus two standard deviation for the difference images. For the multitemporal regression, pixels were considered as burnt if $NDVI_A < (NDVI_{AR} - 2S)$, where $NDVI_A$ is the observed NDVI value after the fire, $NDVI_{AR}$ is the estimated NDVI value after the fire, and S is the regression error within each window. The multitemporal regression approach performed slightly better than the image differencing, and both methods produced unbiased and accurate estimates of the area of large fires.

Cahoon et al. (1994) analysed the severe 1987 fires in northern China and south-eastern Siberia with AVHRR Global Area Coverage (GAC) imagery. They created a clear-sky mosaic with images from multiple dates, and composited them using minimisation of channel 2 reflectance as the compositing criterion. The extent of the burnt area was then assessed by classifying the mosaic with the minimum distance to means classifier, applied to AVHRR channels 1, 2 and 4.

Burnt area mapping with a supervised classification procedure was performed by Pereira et al. (1998), who used the Classification And Regression Trees (CART) algorithm of Breiman et al. (1984) to induce a classification tree from training data. They combined imagery from the Iberia Peninsula and from central Africa, and used CART to induce a single set of rules capable of segmenting NOAA-AVHRR imagery from the two regions, into burnt surface, unburnt surface and clouds. Classification accuracy, assessed from independent training data, was very high, and comparison with fire perimeters derived from Landsat TM imagery (for Iberia), and with active fires (for Africa), also revealed a very good match.

Caetano et al. (1996) used a single NOAA/AVHRR 1.1 km image from mid-September of 1991 to detect and map fires occurring during that Summer season in central Portugal. They performed a spectral unmixing analysis, based on image endmembers, and used channel 1 (red), channel 2 (NIR), and NDVI spectral data. Green vegetation, soil, and burnt surface were the three spectral endmembers used in the SMA. Underflows and overflows (i.e. endmember fractions lower than zero, and higher than one, respectively) were accepted, and used as model fitting diagnostics. Iterative analysis of endmember fraction and RMSE images served to optimise the selection of image endmembers. The resulting burnt fraction image was density sliced to define three classes: burnt, partially burnt, and unburnt. Partially burnt pixels adjacent to burnt ones were reclassified to burnt. AVHRR-based areal estimates of burnt areas were compared against a high spatial resolution fire perimeters map, developed from Landsat TM imagery. The spectral unmixing approach, complemented by the spatial adjacency criterion led to a correct classification of 91% of the burnt area, but at the cost of a relatively large level of commission error, mostly due to areas where fires from the 1991 season were adjacent to 1990 burns. Spectral unmixing was considered advantageous over VI-based methods, due to its improved capability to distinguish burns from other unvegetated or sparsely vegetated areas.

Eva et al. (1995) analysed fires in the savannas of central Africa with a time series of data from the first Along-Track Scanning Radiometer (ATSR-1). They considered that a burnt area was present when a sharp decrease in reflectance at the 1.6 μm channel was accompanied by an increase in brightness temperature at the surface, given that the magnitude of both of these changes exceeded certain thresholds. The 1 km spatial resolution of ATSR-1 was considered adequate to estimate areas burnt in open savannas, but less reliable in forest-savanna mosaics, where fires tend to be smaller.

Barbosa et al. (1997, 1998b) also employed a time series approach, but worked with weekly composites of NOAA-AVHRR data, using minimum albedo as the multitemporal compositing criterion (Barbosa et al. 1998a). They used Global Area Coverage (GAC) data at 5 km resolution to analyse a multiple year data set of Africa and relied on surface brightness temperature and the GEMI to detect burns. A fire was considered to have occurred whenever a decrease in GEMI took place simultaneously with an increase in temperature. The thresholds for such a change to be considered as representing a burn were automatically adjusted, based on the mean and standard deviation values of the time series.

Multiple-thresholding approaches, similar to those previously used for active fire detection (Kaufman et al. 1990; Kennedy et al. 1994) were employed by Pereira et al. (1998) and Sousa (1998). The main difference is that traditional crisp thresholds were replaced by fuzzy thresholds, or membership functions, derived through visual inspection of the imagery. Pereira et al. (1998a) mapped daily burnt areas in a series of 25 NOAA AVHRR images of central Africa, relying on the GEMI3, albedo, and channel 3 reflectance. Comparison of daily estimates of area burnt with the occurrence of active fires detected with the algorithm of Flasse and Cecatto (1996) revealed very good agreement between the timing and general location of scars and active fires, but the latter severely underestimate area burnt. Sousa (1998) used a similar approach, but relying on GEMI, albedo, and AVHRR channel 4 brightness temperature. Comparison of her results with fire perimeters derived from Landsat TM showed good agreement (r = 0.90) and low regression bias.

In the remainder of this chapter, we present results from our research on burnt area mapping in southern Europe, using imagery from the NOAA-AVHRR. Our analyses concentrate on the Iberian Peninsula, the area in southern Europe where fire incidence is highest (CE 1996).

8.3
Mapping burnt areas in southern Europe from NOAA-AVHRR data

8.3.1
Data and methods

The study area is located between the latitudes of 34° 20' N and 45° 50' N, and the longitudes of 24° 40' E and 11° 42' W. It covers the entire territories of Portugal and Spain (with the exception of the Azores, Madeira and Canary Islands), Italy, Greece, and the southern part of France. The data used in the project were acquired in High Resolution Picture Transmission (HRPT) format by the NOAA-11 satellite, during the early afternoon pass. The images were navigated, reprojected to Albers Conical Equal Area (ACEA) projection, and geometrically adjusted using ground control points, mostly from the extensive coastlines of southern Europe. The data were also radiometrically calibrated to apparent reflectances (channels 1 and 2), taking into account the calibration drift (Rao and Chen 1995), and to brightness temperatures (channels 4 and 5), correcting for non-linearity

(Cracknell 1997). Due to an idiosincracy of the archiving procedure, channel 3 data were unavailable for the period of the study.

Two distinct methodological approaches were followed for the years of 1991 and 1994, which were record years for burnt area in Portugal and Spain, respectively. A common step in both approaches was multitemporal compositing of the daily AVHRR imagery, to reduce cloud coverage and radiometric instability of the data due to other atmospheric and directional effects. New compositing criteria had to be developed, because the traditional criterion of compositing according to the NDVI MVC is inappropriate for burnt area analyses (Barbosa et al. 1998; Pereira et al. 1998). However, the alternative approaches proposed by Cahoon et al. (1994), based on minimum channel 2 compositing, and by Barbosa et al. (1998), using minimum albedo compositing, were also found to be unsuitable due to retention of cloud shadows over burnt areas. Therefore, a new two-criteria procedure was applied to the 1991 imagery, whereby the three dates with the lower channel 2 reflectances in each 10-day compositing period are kept, for each pixel. Then, the date out of these three with the highest channel 4 brightness temperature is selected (minC2 → maxC4). For the 1994 imagery, a simpler approach based on maximum channel 4 brightness temperature was deemed adequate. Both approaches tend to preserve the signal from recent burns over practically any alternative, and do not retain cloud shadows in the composited images of the individual channels.

The two approaches diverged in subsequent steps. The main phases of the procedure used to analyse the 1991 imagery are summarised in Fig. 8.1. Eleven dates of pre-processed AVHRR imagery were composited as described above, and used to calculate the Global Environmental Monitoring Index (GEMI) of Pinty and Verstraete (1992), albedo (Saunders 1990), and brightness temperature, according to the split-window procedure of Price (1984). These variables are sensitive to the main spectral properties of recently burnt surfaces, and their usefulness for burnt area mapping has been demonstrated by Pereira et al. (1998a,b) and Sousa (1998). Training data from burnt areas, and from the unburnt landscape were extracted by visual inspection of colour composite images, and used in a supervised classification procedure. We used classification and regression trees (CART), a non-linear, non-parametric classifier that generates classification rules through an induction procedure described in detail by Breiman et al. (1984). CART fits models by a procedure of binary recursive partitioning, that successively splits the data into increasingly homogeneous subsets. Results are presented as an inverted tree, starting in a root node, and generating descendent nodes through series of yes/no questions. Some nodes are terminal, indicating that a final classification was reached, while other (child nodes) require further splitting, until a terminal node is reached. Each split separates parent node into exactly two child nodes, and is usually based on a question relating to a single variable. CART was successfully used by Pereira et al. (1998b) to develop a single classification tree capable of simultaneously mapping burns in Iberia and the Central African Republic.

Accuracy of the resulting burnt area maps was assessed in two ways. In the first, we used a 10-fold cross validation procedure (Breiman et al. 1984). The data were divided into ten equal sized subsets, and ten different classification trees were grown, each one using 9/10 of the data to develop the model, and the remaining 1/10 for accuracy assessment purposes. Then, the tree which optimises a

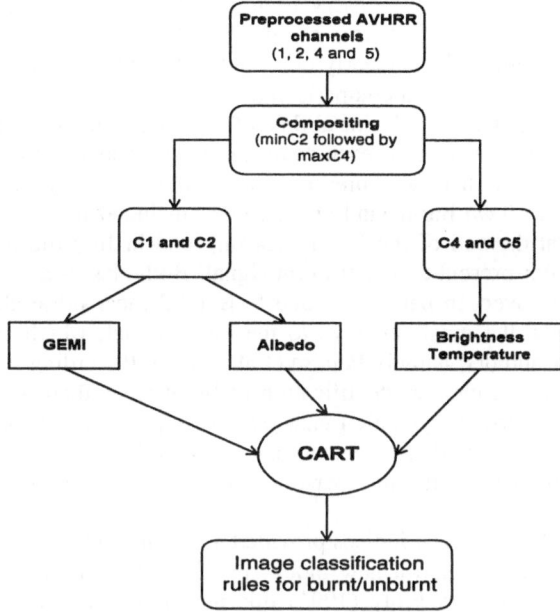

Fig. 8.1. Image analysis flowchart for the 1991 AVHRR data set.

compromise between simplicity of the rule set and classification accuracy statistics (the optimal tree) was selected. Classification accuracy assessment was performed with a contingency table and the κ (kappa) statistic. The second procedure compared area estimates for fires larger than 500 ha, obtained with Landsat 5 TM data, and with the CART-based AVHRR image classification. Sousa et al. (1997) describe the production of the high resolution TM burnt area map. This procedure was applied to the fires that occurred in Portugal during the summer of 1991, and also to a small number of large fires from Andalucía and Valencia, in Spain.

For the 1994 image data once the multitemporal registration was solved with a reasonable level of accuracy, a radiometric analysis was undertaken in order to identify the more sensitive bands for burnt land discrimination. According to the conclusions of Pereira et al. (1997), this analysis included original bands 1 and 2, as well as derived variables such as NDVI, SAVI, GEMI, linear mixture analysis and surface temperature. Several training fields of burnt areas and other relevant land covers were extracted from selected 1994 images. Previous cloud masking was undertaking, in order to avoid confusions. Average values were computed for the whole series of images (31, from June to September).

Periodic noise introduced in AVHRR images by the orbital drift was very obvious. Average values computed from each image (after cloud masking) tend to present a cyclical pattern, related to the daily observation geometry. Since the NOAA satellite repeats the same observation geometry every 9 days, daily coverage from the AVHRR sensor requires a wide scan angle. Consequently, any target area is observed from very diverse angles in subsequent days. In principle, the more off-nadir, the noisier the observation. To prevent the use of spurious data, AVHRR mosaics were created from acquisition angles lower than 30°. However, the effect of this daily drift is clearly observed in the daily average values, which will complicate multitemporal comparison between images.

This tendency towards daily instability of AVHRR values is also clearly observed in the average values of each cover class, especially in the original bands (Plate 8.1). Derived indices also present such a daily trend, but the values are less noisy (Plate 8.2). Two burnt land classes were included in this analysis: pixels burnt in June (at the end of the Spring season), and in July (during the Summer season). The latter present a clearer burnt signal. Both classes are not distinguishable from other covers in bands 1, 4 and 5. Band 2 (near infrared) is among the original channels, the most sensitive to the burnt signal, which agrees with the finding of other authors, namely Pereira et al. (1997). Regarding the derived indices, GEMI permits a clearer identification of burnt areas than NDVI and SAVI, but it also had the lower dynamic range. This is a quite logical finding, since the GEMI is less sensitive to dense vegetation cover than NDVI, and it is more appropriate to discriminate soils and sparsely vegetated areas (Pinty and Verstraete 1992).

Since none of the derived indices provided a complete clear separation of burnt areas, we decided to create a new index specifically designed for this purpose. Following Verstraete and Pinty (1996) this new index tended to minimise the distance from a so called "convergence point", defined as the extreme spectral value of burnt areas in channels 1 and 2. The former was selected because it provides a good estimation of background albedo, while the latter is very sensitive to burnt areas. Therefore, the Burnt Area Index (BAI) was defined as follows:

$$BAI = \frac{1}{(\rho c_r - \rho_r)^2 + (\rho c_{ir} - \rho_{ir})^2} \; , \qquad (1)$$

where ρc_r and ρc_{ir} are the convergence values of red and near infrared bands, respectively. Since the BAI is computed as the inverse squared distance from the convergence points, the values will tend to be higher when closer to the convergence point, and lower in areas with spectral values far from it. Therefore, the values of the convergence point are critical to obtain proper results. The convergence point was defined based on a previous literature review and analysis of AVHRR images from 1991, 1994 and 1995, as 0.1 for ρc_r and 0.06 for ρc_{ir}. The main problem with AVHRR images refers to the lack of radiometric consistency in the temporal series. Consequently, in theory, proper working of the BAI would require either daily change of the convergence point, or modification of BAI thresholds for burnt areas.

Once the most appropriate variables for burnt area discrimination were defined, the above mentioned compositing technique based on channel 4 brightness temperatures was applied obtaining two composites: one with the images of July and August and the other from August and September images. Afterwards, a thresholding criteria was applied in order to discriminate between burnt and not burnt areas. Most commonly, these thresholds are based on the mean and standard deviation, assuming they properly describe each variable's central tendency and dispersion. For the purposes of burnt land mapping, thresholds were established using the mean plus 7 standard deviations, in the case of the BAI, and the mean plus 2 standard deviations in the case of channel 2 and GEMI. As mentioned before, the former is more sensitive to burnt areas and provides a much wider range than channel 2 or GEMI. Consequently, the thresholds could be more restrictive to avoid commission errors.

Since burnt pixels are potentially confounded with other land covers, such as urban areas, shaded forests, and sparsely vegetated soils (Chuvieco and Congalton 1988), burnt land mapping was accomplished in two phases. The first one was meant to identify burnt pixels within a scorched area, reducing commission errors as much as possible. Once these pixels were targeted using the single thresholding methodology described above, the second phase focused on deriving the whole area affected by the fire, considering only the neighbourhood of those "core" burnt-pixels. To obtain homogeneous regions from starting pixels requires an auxiliary variable, from which previously detected pixels can be extended to the whole affected area. Obviously, this auxiliary variable should properly enhance the spectral signal of burnt areas. As said before, channel 2, GEMI and BAI were able to discriminate burnt areas from other covers, but since a large fire includes different levels of scorching and mixtures of vegetation and soils, those variables presented high variances (and therefore low homogeneity) that precluded the application of region-growing algorithms. Instead, we decided to use spectral mixture analysis (SMA), which as previously mentioned it is a very sensitive technique to estimate the proportion of burnt area in each pixel. Since only three bands were available, the number of pure classes should be restricted to two, in our case healthy and burnt vegetation. Additionally, an error band was derived to measure the model fitting. The spectral unmixing of the burnt-land class provided a good image of the affected areas (Fig. 8.2 shows this component for the Iberian Peninsula).

Initially, we planned to use region-growing algorithms to obtain fire perimeters from pixels previously discriminated as fires. These techniques only require a "seed" pixel (or a set of pixels), from which homogeneous measures are computed to include or not surrounding pixels. Those homogeneous measurements would be computed from the burnt proportions derived from SMA.

Unfortunately, the algorithms tested did not work properly. Therefore, we used instead a new criterion based on the spectral and spatial distances from the pixels previously detected as burnt. This approach is based on the concept of distance from the ideal point, commonly used in multicriteria evaluation (MCE) techniques (Zeleny 1982). The basis of this method is to establish the coordinates of a point that is assumed to be ideally located in a certain measurement space. All other points are classified in levels of suitability according to the proximity to the ideal point. Conversely, the distance to the anti-ideal point can be computed. In this case, the farther the distance, the higher the level of suitability. In our case, the coordinates of the ideal point were defined from pixels previously discriminated as burnt areas. The spectral coordinate was the average of affected pixels in the burnt component of the SMA, while the spatial coordinate was defined as the minimum distance to those fire-affected pixels (that is zero). Therefore, the distance to the ideal point (d_{ip}) was defined as:

Fig. 8.2. Burnt-land component of the Spectral Mixture Analysis for the Iberian Peninsula. August-september 1994. Fire affected areas shown in white.

$$d_{ip} = \sqrt{(INC_i - INC_m)^2 + DIS_i{}^2} ,$$ (2)

where INC_i is the value of the burnt component of SMA for pixel i; INC_m the average value of that component for all the pixels previously discriminated as burnt areas; and DIS_i the distance of pixel i to the closest pixel discriminated as burnt.

Similarly, the distance to the anti-ideal point (d_{aip}) was defined as:

$$d_{aip} = \sqrt{(INC_i - INC_{nq})^2 + (DIS_i - lim)^2} ,$$ (3)

where INC_{nq} is the average value of the burnt component of the SMA for pixels previously discriminated as not affected, and *lim* indicates a threshold of distance from which the likelihood of a pixel being included in any fire perimeter should be close to zero.

8.3.2
Results

The multitemporal compositing procedure applied to the 1991 data set (minC2 → maxC4) was quite effective in removing clouds from the imagery, both over land and over the ocean. This is apparent in a red-green-blue (RGB) colour composite for the whole study area, using AVHRR channels 4, 2, and 1, respectively (Plate 8.3). In this image, the green colour represents densely vegetated areas, and the red, white, and bluish tones indicate sparsely vegetated areas, with variable colour soil backgrounds. Burn scars, barely visible at the scale of the figure, appear bright red. A cluster of burns is visible in central Portugal, but the large red patch along the southern bank of the Guadalquivir river in southwestern Spain, although spectrally very similar, is unlikely to be a burn, due to its very large size and location in a predominantly agricultural area. This example illustrates some of the difficulties in burnt area detection and mapping.

Plate 8.4 shows an RGB colour composite of the Iberian Peninsula, using brightness temperature (red), GEMI (green), and albedo (blue). In this composition, green is again representative of vegetation, with darker hues for forested areas, and brighter hues for irrigated agriculture, especially evident along the Ebro and Guadalquivir valleys, in northeastern and southwestern Spain, respectively. Sparsely vegetated to bare soils appear in shades of red, orange and purple. The latter colour indicates bright, hot soils, while the purer red hue represents hot, darker soils. The cluster of fires in central Portugal is again visible here, as well as some very large fires along the Mediterranean coast of Spain, all coloured bright red. A large dark smoke plume can be seen near the latter cluster of burns, oriented in a southwest-northeast direction. Large patches of landscape spectrally similar to burns in this colour composite are again visible throughout southern Spain.

The classification tree, or set of rules produced by CART through induction from training data is represented in Fig. 8.3. Non-terminal nodes are shown as white diamonds, and terminal nodes as grey squares. The path down from the root node (the one on top), down the tree to a terminal node, specifies a classification rule. Class 0 indicates an unburnt pixel, and class 1 a burnt pixel. Each non-terminal node contains an identification of the variable selected for splitting at that level, the value of the threshold at which the splitting was performed, and the number of pixels affected by that decision. All pixels with values lower or equal to the threshold go to the left side of the node, and all others to the right side. GEMI was the variable selected to perform the main split, basically segmenting the training data set into densely vegetated pixels (GEMI > 0.382), and less vegetated pixels. The former are immediately allocated to the unburnt class. The main classification rule for burnt surfaces is:

If GEMI <= 0.382 AND Surface Temperature > 312.83 K AND albedo <= 0.108 AND albedo <= 0.096 THEN class = 1.

This rule illustrates a feature of CART, which is the ability to use the same variable multiple times in the same rule, in this case albedo. If minimisation is used as the logical operator to implement set intersection, then the second (lower) albedo threshold supersedes the first one.

Figure 8.4a displays, for Iberia, the results of segmenting the southern European 1991 dataset with the CART-induced tree. Figures 8.4b and 8.4c zoom in on the two areas identified in Fig. 8.4a. In the central Portugal zoom, burn scars (light grey) were overlaid with the fire perimeters derived from Landsat TM imagery,

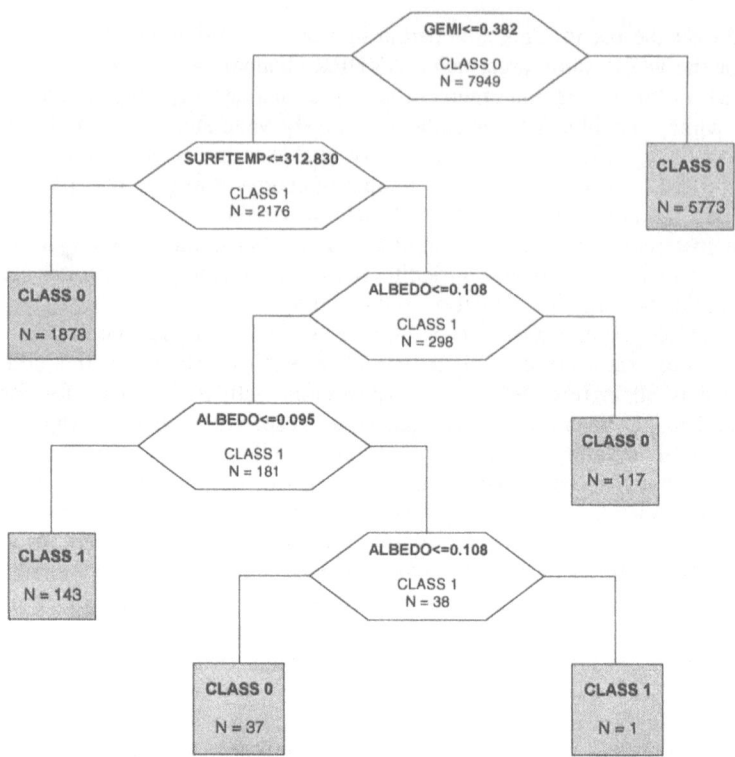

Fig. 8.3. Tree-structured rule set produced by CART through induction from training data. For the terminal nodes, class 0 codes unburnt pixels, and class 1 burnt pixels.

and reveal a very good match. The large patches in Spain, just east of the border, are actual burns, but for which high spatial resolution data were unavailable. However, visual inspection of the colour composites clearly indicates that they were correctly classified. In the Andalucía zoom, fire perimeter data were available only for some of the burns. However, the only area we firmly believe is a false alarm is the one on the lower right corner of the window, which belongs to the already mentioned dryland farming region south of the Guadalquivir river. In the opposite case, of a high resolution fire perimeter that has no corresponding burn identified in the segmented image, we believe it may have occurred at a later date, since the area appears quite green in both colour composites.

Classification accuracy statistics, based on a 10-fold cross validation procedure (Breiman et al. 1984) are shown in Table 8.1. The accuracy is very high, but one must keep in mind that the data were sufficiently prototypical of their respective classes to have been selected as potential training candidates, and this is likely to inflate accuracy assessment figures (Stehman and Czaplewski 1998). A more stringent assessment of burnt area classification accuracy was derived by correlating the areas of all individual fires larger than 500 ha, as estimated with Landsat TM imagery, and with the current AVHRR image classification procedure. Due to

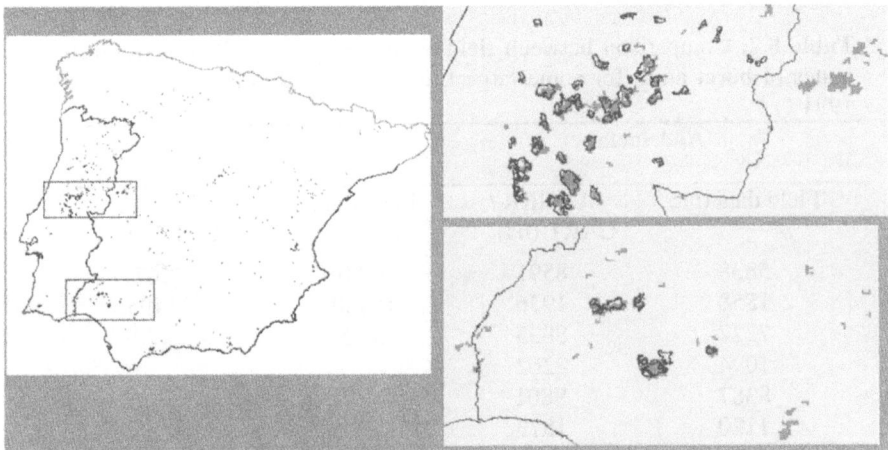

Fig. 8.4. a) Segmentation of the multitemporal composite data for Iberia, using the rules of figure 8.3. b) Detail of 8.4a over central Portugal. Fire perimeters derived from Landsat 5 TM data are overlaid on the areas classified as burnt. No TM data were available for the burns in Spain. c) Detail of 8.4a over southern Spain. High resolution fire perimeter data were not available for all burns. See text for further discussion.

Table 8.1. Contingency matrix showing classification accuracy statistics for the rule-based classification developed with CART

		Predicted Class			
		unburnt	Burnt	Total	Omission Error
Actual	unburnt	7804	11	7815	0.00141
Class	burnt	3	131	134	0.02239
	Total	7807	142	7949	
Commission Error		0.00038	0.07746		

Kappa = 0.95.

the need for exhaustive fire perimeter data, this was done only for those fires that occurred in Portugal, during the summer of 1991.

Results of a regression between the AVHRR and Landsat TM data are depicted in Fig. 8.5. This analysis shows that there exists a very good correlation between the two area estimates ($r = 0.91$) and also that the regression equation is essentially unbiased. The slope of the regression equation is very close to 1 and the intercept is not significantly different from 0 ($p = 0.000$). The standard error of the estimate is of about eight AVHRR pixels.

Table 8.2 compares field and AVHRR area estimates for some large fires in Spain. Lack of exhaustive high spatial resolution data precludes a more thorough and quantitative assessment of classifier performance in Spain, but this informal comparison also reveals good agreement.

As previously mentioned, in 1994 burnt land mapping was undertaken in two phases. The first one was focused on discriminating those pixels most likely to have been affected by a large fire, and the second one on the accurate delineation of burnt areas. Since the former was addressed to reduce commission errors as

Table 8.2. Comparison between field estimates and AVHRR/CART estimates of burnt areas for some large fires in Spain, during the summer of 1991

Andalucía		Valencia	
Field data (ha)	AVHRR / CART (ha)	Field data (ha)	AVHRR / CART (ha)
5838	8591	3550	3267
1858	1936	15400	11495
7222	8833	3035	2420
1094	2262		
8387	9801		
1120	1573		

much as possible, only the most clearly burnt areas were identified. The second phase should complete the mapping of the fire, starting from those "core" pixels.

For the discrimination phase, several thresholding techniques were applied to NDVI, GEMI and BAI images derived from temporal compositing of 15-20 images taken in July, August and September. As previously mentioned, values of the composites were extracted from the day when the maximum temperature of channel 4 was registered. The method was calibrated for Spain, where an accurate database of large fires was available. Assessment of different methods was per-

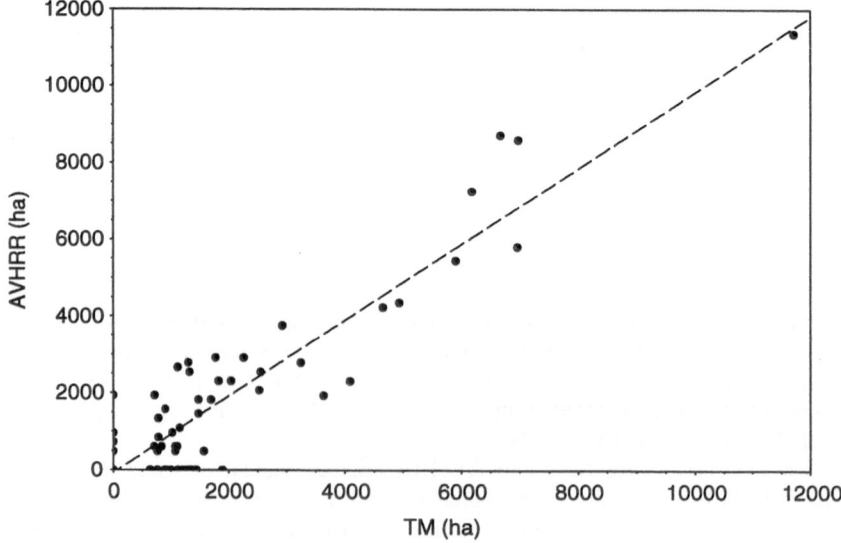

Fig. 8.5. Regression analysis scatterplot, with AVHRR-based fire size estimates shown as a function of TM-based size estimates. The dashed line shown is the regression line ($y = -77.72 + 0.99x$). Standard error of the estimate is 986 ha.

formed by comparing omission and commission errors. The former were defined as those large fires not detected from the images, and the latter the percentage of pixels classified as fire that were not actually burnt. Commission errors are more difficult to assess, since fire statistics do not include a precise mapping of burnt area, but only the location of the 10x10 km grid cell where the fire was detected. However, a large fire may obviously affect areas outside a cell and therefore those pixels cannot with certainty be discarded from belonging to the fire.

After the application of the simple thresholding criterion previously commented to the BAI, GEMI and NDVI images, the highest accuracy was found with the first index. However, several sources of noise were detected in low vegetated areas and water bodies. To avoid the former, the NDVI-MVC image of September of the previous year was used to extract only forested regions. Water bodies were removed from further analysis by a mask created from the Corine Land Cover map.

Table 8.3 includes the accuracy assessment of applying thresholds to the BAI images of Spain for 1994, after applying water and forested mask. As can be noticed in the table, the method is more reliable as the fire size increases. In total, only six large fires were missed, either by poor quality of the images or because they burnt early in the fire season (end of Spring) and therefore the burnt signal was not strong enough to be discriminated.

As far as the mapping phase is concerned, the method of multicriteria distances was applied. In order to simplify the definition of thresholds, both distances to the ideal and anti-ideal point were computed, and a pixel classified as being part of a large fire when the former was greater than the latter. The limit of distances for the anti-ideal point was set to 15 km, since large fires in Spain in 1994 affected vast areas. Lower values would have implied a more precise definition of pixels closer to those previously discriminated as being burnt.

A final map of large fires affecting the Iberian Peninsula in 1994 can be observed in Fig. 8.6, with provincial limits for Spain overlaid. It is obvious that large fires severely affected the Mediterranean regions of Spain and Central Portugal in the study period. In 1994 Spain suffered the worst fire season in the decade. A total area of 251.438 hectares were burnt in 57 large fires, according to our calculations, which do not include the final weeks of August and September. The most affected regions are again the Mediterranean areas, especially Valencia, Castellón, Barcelona and Murcia.

8.4.
Discussion and conclusions

The approach used to map areas burnt in Iberia during 1991 appears to have worked effectively. The new compositing procedure developed for this study generated substantially cloud-free images, while simultaneously enhancing the burnt areas signal, and avoiding the retention of cloud shadowed pixels. This minimises the problems previously mentioned by Cahoon et al. (1994), Barbosa et al. (1998), Martín (1998), and Sousa (1998) regarding multitemporal compositing of AVHRR imagery for burnt area analyses.

Fig. 8.6. Large forest fires in the Iberian Peninsula (1994).

Table 8.3. Large fires detected by multiple thresholding of BAI images in Spain (1994)

Size class (ha)	Fire statistics	NOAA images	
	# Fires (A)	# Fires (B)	(B/A)*100
500 – 1,000	7	3	42.8
1,000 – 2,000	8	7	87.5
2,000 – 3,000	4	3	75
3,000 – 4,000	2	2	100
> 4,000	14	14	100
TOTAL	35	29	82,8
Commission errors (*% pixels improperly identified as fires*)		20	

The spectral indices used for classification, selected on the basis of previous experience (Martín 1998; Sousa 1998; Pereira 1999) performed well, and the tree-structured rule-based classifier developed with the CART algorithm was able to identify the distinctive spectral characteristics of recently burnt surfaces. Both types of accuracy assessment, the one based on independent training data, and the one relying on correlation of areal estimates, yielded very good results.

An interesting aspect of the regression analysis is the observation that, below a threshold of approximately 2000 ha, there are several fires mapped with the TM data that go undetected in the AVHRR image classification, and also quite a few false alarms, i.e. areas classified as burnt by the CART-based model, that do not correspond to a fire perimeter mapped with the TM. We believe these results indicate that the detectability threshold for vegetation fires, with AVHRR 1.1 km data, in southern European landscapes, should reside somewhere between about 1,000 ha and 2,000 ha. This is confirmed by the value of the standard error of the estimate obtained in the regression analysis. These results are, however, contingent on data from a single fire season, and rigorously assessed over only a small part of the entire study area. It is important, therefore, to confirm them with further similar analyses, which are already in progress. The DEF/ISA team is exploring the feasibility of improving the accuracy of the estimates for burns in the lower size range, using image fusion techniques to combine AVHRR data with imagery from the 180 m spatial resolution Wide Field-of-View sensor (WiFS), installed aboard the Indian Remote Sensing satellite (IRS). We are also trying to improve the robustness of the classification trees induced from the data, so that they may work effectively with various data sets from different dates. This is being attempted with classification trees that use fuzzy thresholds, according to the algorithm of Janikow (1998).

The methodology applied to the 1994 data set has also produced good results for the discrimination of fires larger than 1000 ha. The proposed BAI can be considered a first step towards the formulation of a specific spectral index adapted to discrimination of burnt signal. However, it would be of interest to test the performance of this index in different environments and using other sensors prior to establishing a definitive conclusion about its suitability.

Acknowledgements. The research reported in this chapter was developed under the Megafires research project (contract ENV4-CT96-0256), funded by DG XII of the European Commission. We also thank the Direcção-Geral das Florestas (Portugal) for allowing us to use their high resolution fire perimeter data.

9   Burnt land mapping at local scale

N. Koutsias, M. Karteris
Department of Forestry and Natural Environment, Aristotelian University (Greece).

A. Fernández-Palacios, C. Navarro, J. Jurado
Consejería de Medio Ambiente, Junta de Andalucía, Sevilla (Spain).

R. Navarro
ETSI Agrónomos y de Montes, Universidad de Córdoba (Spain).

and A. Lobo
Institut de Ciències de la Terra, Barcelona (Spain).

Abstract. This chapter outlines some of the aspects of the contribution of satellite remote sensing, especially when high spatial resolution data are involved, to burnt land studies. Some theoretical aspects of the meaning of scale and how it influences these studies are also provided. A brief description of the particular character of the Mediterranean landscape is given, because it constitutes the geographical domain to which this book refers. Finally, a description of some of the techniques that have been developed and used in burnt land studies (mainly from the Megafires project) is provided.

9.1
Introduction

Wildland fire occurrence is a critical component for various ecosystem types spread around the whole world, such as boreal forests, temperate forests, Mediterranean ecosystems, grasslands and savannas, vegetation in arid and semi-arid regions, tropic and exotic plantations, etc. However, it reacts in different ways in each one of them and is also associated with different reasons and effects (Chandler et al. 1983). Also, forest fires often constitute a powerful land management tool that is used as a short or long-term landscape modifier (Salvador and Pons 1995, Pereira et al. 1997).

In the Mediterranean European Basin, which is partially covered by productive and potentially productive forests, sclerophyllus shrublands (Maquis) and grasslands, forest fires play a fundamental role. The forests and shrublands, especially those growing in low altitudes, where the climate is characterised as typical Mediterranean, are well adapted in extreme climatic conditions, such as droughts (Karteris 1995). The particular characteristics of the Mediterranean-type vegetation, which has been formed through long adaptation mechanisms under these conditions (i.e. fire-dependent ecosystems, flammable vegetation types), plus the Mediterranean-type climate characterised by strong winds and prolonged dry summers, favour fire occurrence and spread. In the Mediterranean Basin, forest wildland fires constitute a major ecological process, which have a profound influence on the natural cycle of vegetation succession and on the ecosystem's structure and function (Koutsias and Karteris 1998a). Forest fires may exert a positive or negative influence upon ecosystem dynamism depending on the particular characteristics of fire occurrence such as intensity, type, periodicity etc. However, the high number of forest fires occurring every year and over the same areas, which in turn amounts to thousands hectares of burnt land in the Mediterranean Basin, constitutes a real threat to natural ecosystems.

The economic, social, ecological, atmospheric and climatic consequences associated with fire activity denote not only the magnitude of the problem, but also impose the development of a comprehensive information system including advanced and powerful monitoring processes. Inventory of the areas affected by wildland fires, as regards their accurate location and mapping of the ignition points and the scorched perimeter, as well as the species affected and severity of damage, is important in order to estimate the economic losses, assess the ecological disturbances, map the land cover/use changes and estimate short and long-term post-fire consequences. The latter include the study of further deterioration of burnt land due to soil erosion and desertification processes, and also the assessment of the atmospheric and climatic impacts on local and global scales (Martín et al. 1994, Karteris 1995, Pereira et al. 1997). On the other hand, the establishment of a rational management plan in order to protect and restore, naturally or artificially, the areas affected by wildland fires to the pre-fire situation, presupposes their accurate location and mapping (Koutsias and Karteris 1998b).

A critical issue which affects fire management is associated with the lack of appropriate statistics in geographic and measurement scales that allow detailed description of fire incidence (Martín et al. 1994). A well-structured decision-making system for the rational management of forest fires requires a complete, detailed and accurate spatial database of burnt areas. Moreover, appropriate statistics on fire incidence on a permanent basis will help fire managers to better understand the fire problem including the reasons for fire occurrence and spreading (Koutsias and Karteris 1998a). On the other hand, the estimation of these statistics should be accomplished by using effective, quick and accurate methods, which will ensure an appropriate inventory and assessment system.

Satellite remote sensing could effectively be involved in burnt land mapping, since it provides the necessary means for gathering information of the Earth's surface in a less expensive and timely fashion. Periodic spectral data in the visible and infrared part of the electromagnetic spectrum, of high spatial resolution, acquired from remote satellite sensors, offer an unlimited basic source of information. Actually these data by appropriate computer assisted processing and interpretation can contribute to a better, cost-effective, objective and time saving method for monitoring and mapping areas affected by wildland fires.

9.2
Scale issues in burnt land mapping

The sense of scale is a critical component for all studies involved in the spatial or geographic domain. Inventory, analysis and rendering of phenomena which occur in the geographic or temporal space comprise the concept of scale as a fundamental component, which determines and/or is determined by the approach level of the phenomenon itself. Scale, which may refer to either the spatial or temporal dimension of the event studied, lends particular characteristics of the methodological approach followed and the expected results as well. For reliable, or even feasible, conclusions about the behaviour of any phenomenon, the study of the phenomenon itself should be carried out within a range of scales (spatial or temporal) that allow accurate observation and function. Minimum and maximum values define this range and determine the operational scale. Outside this range the phenomenon is not observable, while within it, the level of scales determines the degree and the accuracy of the study.

Lam and Quatrochi (1992) considered that the term scale has a variety of meanings depending on the context used. It may refer either to space (spatial scale) or to time (temporal scale) as well as to a combination of both (spatio-temporal scale). Actually, inside each of these three different aspects of scale, there are different connotations depending on the particular characteristics to which the scale refers. According to Bian (1997), Lam and Quatrochi distinguished the spatial scale into four connotations: cartographic, geographic, operational and resolution, although the concept of resolution was considered to be closely related to the others. Moreover, Cao and Lam (1997) used the term "measurement", instead of resolution, to define the fourth connotation. For burnt land mapping, the classification proposed by Cao and Lam (1997) was adopted and the above four meanings are further discussed and examined.

The first, cartographic or map scale, is the traditional meaning of scale refering to the scale of cartographic or map products and it is estimated by the ratio of a distance in map units to the corresponding distance on the ground. A large-scale map may depict a small burnt area and consequently gives more detailed information about it (Plate 9.1). When the purpose of the cartographic product is to depict detailed information about the characteristics, spatial and descriptive, of a burnt area, then it presupposes a large cartographic or map scale. On the other hand, in a map which aims to depict the spatial distribution of forest fires, in order to explore the spatial pattern on a national or regional level, a small-scale map is more appropriate.

The second, geographic or observational scale, is used in the context of the spatial extent of the occurrence of the phenomenon studied. A large geographic scale map depicts a large geographic area to which burnt land mapping may refer (Plate 9.1). It is associated and determines mainly what methods and tools are more appropriate to be used for the monitoring and mapping of burnt areas. If the geographic scale refers to a national or regional scale, remote sensing may be more suitable than field survey.

The third, operational scale, which is used in the context of the operational space, denotes the hierarchical level at which a phenomenon occurs. Study of post-fire consequences on a tree level are quite different from those on the forest or landscape level, since landscape operates at a larger scale than the forest and the forest operates at a larger scale than a single tree (Plate 9.1). Depending on the aim of the study, an appropriate scale will ensure its success.

Finally the fourth connotation, measurement scale, which refers to the spatial resolution related to the size of the smallest discrete distinguishable object, reflects the degree of detail and is directly associated with remote sensing studies (Plate 9.1). For some cases, in remote sensing applications, the actual resolution could be smaller or larger than the theoretical one given by the operational characteristics of the sensor, depending on the spatial arrangement of the objects. High geometric resolution gives more detailed information. Large-coverage studies, such as those referring to a regional or national level, utilise coarse resolution satellite data. In doing so, they assure their feasibility as regards the cost of data acquisition and storage and processing requirements (Lam and Quattrochi 1992, Bian 1997, Cao and Lam 1997, Goodchild and Quattrochi 1997).

All the above mentioned connotations of spatial scale are closely related, as each of them has a direct or indirect impact on the other. Thus, in remote sensing studies a measurement scale of 1100 m, which may correspond to NOAA-AVHRR data, determines also the cartographic, geographic and operational scale of the study. This coarse resolution of the satellite data usually corresponds to projects referring to continental, regional or national levels, since they are associated with small cartographic scales and large geographic and operational scales. Instead, a measurement scale of 30 m which may arise from satellite data of Landsat Thematic Mapper (TM) corresponds to detailed cartographic scales and small geographic and operational scales and in general refers to the local level.

Closely related to the above four connotations of the spatial scale is cost, another critical component which, in many cases, prohibits the use of such studies. Detailed data, which cover large-scale studies, are expensive both to collect and to process. As a general rule, a balance between detail, accuracy, feasibility and cost makes the application of such studies possible.

9.3
Operational burnt land mapping in Mediterranean landscapes

9.3.1
Structure of the Mediterranean landscape

One of the most particular characteristics of vegetation communities and landscape structure in the Mediterranean Basin is the high diversity and heterogeneity, in terms of the spatial distribution and arrangement of the abiotic, biotic, floristic and ecological components. Giving a short description, Forman and Gordon (1986) suggested that the Mediterranean landscape is composed of a "mixture of large, small, distinct and indistinct patches" (Plate 9.2). This complex spatial nature of the Mediterranean landscape raises a number of questions associated with the optimum scale of measurements that must be made to represent accurately, in the spatial domain, the phenomenon under study.

On the other hand, the irregular terrain relief found across the Mediterranean Basin, with high elevations and steep slopes, acts as another source of distortion in the quality of the spectral information provided by satellite sensors (Plate 9.2). Projects carried out in mountainous areas with irregular relief reported difficulties in discriminating burnt from unburnt land located in shadowed areas (Tanaka et al. 1983, Milne 1986, Chuvieco and Congalton 1988).

The defragmented landscape pattern constitutes the weak point of the applied remote sensing, because the huge amount of interchanges in the landscape form and its complexity make the interpretation and processing of the satellite data difficult. This special kind of pattern, compared to the technical characteristics of remotely sensed data, affects the minimum mapping unit, which defines the smallest distinguishable object to identify and analyse (Goodchild and Quattrochi 1997). Especially in cases when the sensor geometric resolution is coarse, the quality and potential discriminator ability of the spectral information are limited.

As has already been mentioned, the four connotations of spatial scale which are directly associated with remote sensing studies determine three critical concepts in operational burnt land mapping; accuracy, cost and feasibility. Accuracy and cost are two interdependent issues, which determine the feasibility of burnt land mapping in an operational fashion. The achievement of high accuracies requires detailed data of high geometric resolution, which are expensive to collect and analyse. On the other hand, given the geographic extent of the study area, which may be on a national level, it is obvious that there is a certain limit that geometric resolution cannot exceed. This minimum geometric resolution depends on the limitations of the data storage, manipulation and cost. However, this cannot be clearly defined. Finally, it has been demonstrated that even small differences in geometric resolution have a significant impact on the estimated landscape parameters (Gluck and Rempel 1995).

9.3.2
Methodological approaches for burnt land mapping

Basically, three major methodological approaches can be distinguished depending on the tools available to identify and map the burnt areas, as well as on the scale of measurement. The first, referred to as a micro-scale approach, is based on field surveys and on-site human-made observations (Fig. 9.1). This is a time-consuming, expensive but highly accurate method, although the feasibility of its being included in an operational national burnt land-mapping project is limited. The recorded observations in the spatial domain, due to time and cost limitations, cover only some general statistics and usually only information about the scorched perimeter is provided (Martín et al. 1994). The high accuracy, combined with the elevated cost and time requirements, makes this method suitable only in very specific (local) situations, in which the measurement scale (resolution, detail) is more significant than the geographic scale (extent of the study area).

Fig. 9.1. Three major methodological approaches can be distinguished depending on the available tools used to identify and map the burnt areas, as well as on the level of measurement scale; field survey, aerial photography and satellite technology.

The second method, which is referred to as a meso-scale approach, employs the use of b/w, colour or infrared aerial photography of an approximate scale ranging from 1:5,000 to 1:50,000 (Fig. 9.1). Aerial photography, a pioneer remote sensing source of data, provides information in relatively high measurement and observational scales, which, it should be mentioned, are not free of errors. According to Ambrosia et al. (1998), aerial photography as a remote sensing source of data was employed quite early in forest fire assessment to map fire damage (Arnold 1951) and to estimate forest losses due to fire and diseases (Johnson and Thomas 1951). Although aerial photography offers the possibility to cover larger geographical areas than human-made measurements and to process data and extract the desired information at less cost, its use in an operational national burnt land mapping project still remains limited. On the other hand, a well-trained and experienced personnel is required to assure reliable and independent use, since it relies mainly on subjective qualitative criteria.

Finally, the third method, which is referred to as the macro-scale approach, uses remotely sensed data acquired by various satellite systems in measurement scales usually ranging from 10 to 1100 meters (Fig 9.1). Given the large geographic extent over which a burnt land mapping project may operate, it is clear that remote sensing provides an ideal means of gathering the required information.

Satellite remote sensing gives the possibility to acquire data on the Earth's surface even in areas with limited accessibility and on a regular and permanent basis. According to Caetano et al. (1994), the potentiality of the involvement of satellite remote sensing technology in burnt land mapping is defined by:

- the acquisition of data that present different spectral reflectance characteristics between burnt and unburnt healthy vegetation (Tanaka et al. 1983), especially in the infrared part of the electromagnetic spectrum;
- the effectiveness of the cost/benefit ratio compared to field measurements or aerial photography, especially in cases of large geographic extent (Lauer and Krumpe 1973);
- the periodical acquisition of the required information (Lee et al. 1977) combined with the synoptic view and the low required time for data acquisition, and
- the digital form of the data with all the accompanying advantages, such as speed and objectivity of data processing (Richards 1996).

Remote sensing studies for burnt land mapping have been conducted using satellite data of either high-resolution sensors, such as Landsat MSS or TM, SPOT etc. or low resolution, such as NOAA-AVHRR (see Chap. 8).

Although a large number of different methodologies have been developed, there is no standard classification procedure applied to remotely sensed data for identification and mapping of burnt areas. They vary according to the specific characteristics of the case study (Karteris 1995), although the spectral, spatial or temporal resolution of the satellite data also determines the type of method used (Pereira et al. 1997).

Although satellite remote sensing provides an advantageous methodological approach to identify, map and monitor the burnt areas compared to other traditional ones, it is not free of errors. Projects carried out under various environmental conditions using different satellite data and techniques reported some confusion between burnt land and other land cover/use categories. There is common agreement among scientists that these problems may be summarised as follows (Chuvieco and Congalton 1988, Caetano et al. 1994, Karteris 1995, Pereira et al. 1997):

- Confusion between burnt land and water bodies (Tanaka 1983, Chuvieco and Congalton 1988, Pereira and Setzer 1993, Lombrana 1995). These studies associate the problem of discriminating between burnt land and water bodies with a different source of confusion. Among them, topographically shadowed areas, recently burnt surfaces, mixed land-water and water-vegetation pixels are some examples where spectral similarities responsible for the confusion have been noticed (Pereira et al. 1997).
- Confusion between burnt land and urban areas (Tanaka 1983, Chuvieco and Congalton 1988, Lombrana 1995). It has been found that spectral similarities occur between artificial surfaces and burnt surfaces composed of a mixture of charcoal and exposed soil characteristics, although they can be eliminated by masking out these well-defined urban areas (Pereira et al. 1997).

- Confusion between burnt land and shadows (Milne 1986, Chuvieco and Congalton 1988). Shadowed areas, occurring as a result either of the irregular terrain found especially in mountainous areas or from clouds, are responsible for the incorrect classification of burnt surfaces. Successful efforts to eliminate this confusion include the application of a multitemporal principal component analysis (Pereira et al. 1997) or the use of spectral mixture analysis (Caetano et al. 1994). A recent work, using the Intensity-Hue-Saturation transformation of a three-channel composite of Landsat-5 Thematic Mapper, also proved useful (Koutsias et al. 1998).
- Confusion between slightly burnt land and unburnt vegetation (Lee et al. 1977, Benson and Briggs 1978, Minick and Shain 1981, Milne 1986). This source of confusion is associated mainly with the problem of mixed pixels. Caetano et al. (1994) stated that many intermediate combinations of burnt, unburnt and soil proportions can be found between totally burnt and unburnt pixels.

9.3.3
Advantages of using high resolution sensors

In most European countries, operational inventory of forest fires and their consequences is accomplished by methods mainly based on field surveys. Although they ensure highly accurate measurements, usually they are not able to provide detailed descriptions about the fire characteristics in the spatial domain due to high cost and time requirements. As a result, fire inventories cover only some general statistics of fire incidence in coarse spatial resolution that do not allow the detailed evaluation of their consequences (Martín et al. 1994, Chuvieco 1995). From a fire management perspective, the lack of these detailed descriptions constitutes a weak point, which has a significant influence on the post-fire management strategies. The assessment of vegetation recovery, of deterioration through the desertification process and soil erosion, of short or long-term consequences on the fauna and flora dynamics and on the local bioclimate, etc., are a few examples which show the importance of these detailed fire inventories (Isaacson et al. 1982, Martín et al. 1994, Chuvieco 1995, Karteris 1995, Pereira et al. 1997). Some of these problems and limitations can be partially overcome by the use of remote sensing technology, especially when high spatial resolution satellite data, such as those acquired from Landsat TM, are involved. In addition, a critical component, which imposes the use of high resolution sensors, is associated with some particular features which characterise most European Mediterranean countries, such as forest structure, fire characteristics, level of scale measurements etc.

As regards the first factor, the complicated and defragmented structure of Mediterranean natural ecosystems, as already mentioned, does not allow their proper understanding and description by low-scale measurements. Instead, finer resolutions may be more convenient. However, it should be kept in mind that there is a certain minimum that cannot be exceeded, first because this will not ensure its operational feasibility and second because very fine resolutions do not allow the study of some phenomena which occur at larger operational scales. Chuvieco (1995) mentioned two reasons associated with the Mediterranean landscape structure, in order to explain why a more detailed evaluation of burnt areas is needed. The first refers to the spatial distribution and the pattern of forest resources, which are scarcer, especially when the recreational role of the forests is considered. The second refers to the close spatial integration of the forested and agricultural land, which makes burnt land estimation difficult.

Regarding the second aspect, forest fires in the Mediterranean Basin play a very specific role for two reasons: first, as a natural phenomenon associated with the continuous existence of certain vegetation types such as fire-dependent ecosystems (Chandler et al. 1983); second, as a land management tool utilised in agricultural and livestock practices for converting forested land into other types (Pereira et al. 1997). However, regardless of reason and purpose, the majority of forest fires in the Mediterranean result from human activities, while natural causes, such as lightning, make only a minor contribution. This particular character of Mediterranean forest fires, especially when they are used as a short or long-term landscape modifier, requires a detailed and accurate assessment of their characteristics to assure a better understanding of the problem and also to improve the protective action.

Finally, regarding the third feature, it should be noticed that the range of scale measurements within which a specific phenomenon occurs also determines the corresponding means to examine this phenomenon. Given a paradigm in the Mediterranean Basin, where forest fires are generally small, the monitoring tools could not be the same as those used to study large fires such as those occurring in Yellowstone Park in 1988, which burnt about 3300 km^2. This area, which equals 2.5% of the whole of Greece corresponds to 12.3% of the total forested land. In comparison, the total burnt land in Greece in 1988, one of the most destructive years in the past decade, was about 1100 km^2.

Among the advantages of using satellite data of high spatial resolution we can distinguish the following:

- Improved estimated overall accuracy of both fire perimeter and fire scar mapping. This is mainly associated with the development of an accurate and detailed database of fire statistics.
- Improved discrimination ability to distinguish and map unburnt areas within the fire perimeter. This is mainly associated with the post-fire land-scape modification in terms of its spatial structure and pattern.
- Improved assessment of fire intensities and levels of vegetation damage. This is mainly associated with the assessment of long-term landscape con-sequences, such as vegetation recovery and further deterioration due to soil erosion.
- Improved discrimination ability to map the species affected. This is mainly associated with the assessment of the ecological consequences and the vegetation succession and recovery.

9.4
Techniques for burnt land mapping

The scope of this section is to outline some of the technical aspects of burnt land mapping with remote sensing technology and to give a general description of some of the techniques that have been developed and applied on a local scale. For the better understanding and presentation of those techniques, a case study from a large forest fire which occurred in Attica, Greece, in the surroundings of Lake Marathon, and burnt about 5500 ha, was considered. A multitemporal data set consisting of the pre- and post-fire satellite image (Plate 9.3) of Landsat-5 Thematic Mapper, was used to further explore these techniques.

9.4.1
Overview

So far, several kinds of image classification techniques have been developed and applied to detect and map the burnt areas. They range from simple ones such as visual interpretation and single channel density, to more complex such as principal component and spectral mixture analysis. However, the detection and mapping of burnt land remains somewhat problematic, because their spectral response presents a diverse and complex pattern in the spatial and temporal domain (Pereira et al. 1997).

Two critical aspects associated with fire occurrence determine the spectral behaviour of burnt areas: the deposition of charcoal as direct result of the burning, and the removal of vegetation. This second aspect may also be caused by other factors besides forest fires, such as cutting, grazing, water stress, diseases, etc. (see Chap. 7). From the Landsat images analysed in this section, a strong decrease in reflectance in the near-infrared band (TM4) is obvious, and a clear increase in the mid-infrared region (TM7). Finally, a superior performance of the infrared region of the spectrum over the visible (Fig. 9.2), for distinguishing burnt areas, is observed, because of the minor modification of the spectral response of the burnt areas in the visible channels compared to the pre-fire situation (Koutsias et al. 1998).

Although a large number of methods and techniques have been developed and used in burnt land mapping, there is no standard procedure which assures its successful application under a wide range of conditions. Procedures vary according to some specific characteristics, such as:

- The particular characteristics of the fire itself such as type, size, pattern, etc., which determine what kind of satellite data are more appropriate to be used in terms of spatial and spectral resolution.
- The particular geophysical and eco-climatic conditions of the broad area around fire extent, such as the affected species, the terrain characteristics, the other land cover/use categories, the vegetation type dominating in the area, etc.
- The available satellite data such as multitemporal or single post-fire data sets, type of sensor used in respect to spectral, spatial and radiometric resolution, etc.
- The specific objectives of the study, such as the outline only of the scorch perimeter, the detailed mapping of the affected area, the mapping of the levels of vegetation damage, etc. which in turn determine what kind of satellite data are more appropriate to be used.

The methods and techniques developed and used for burnt land mapping, depending on some particular characteristics, can be discriminated according to different criteria: first, regarding the number of images: use of multitemporal

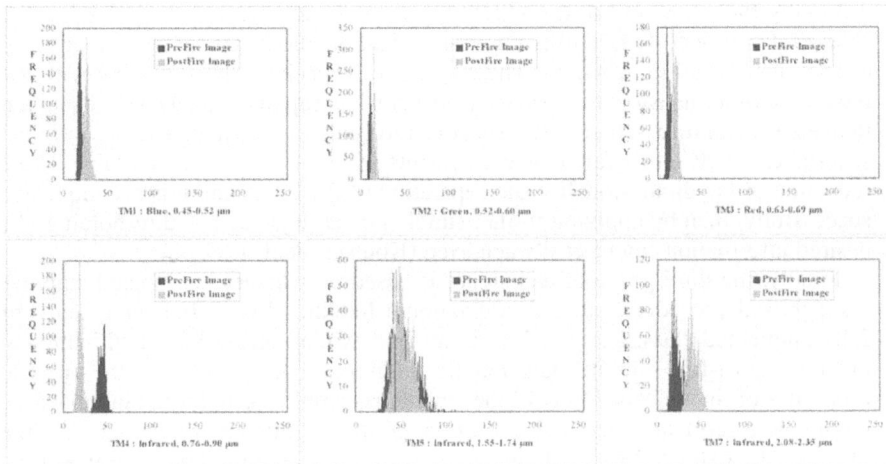

Fig. 9.2. Histogram data plot of a random sample taken from a burnt surface. The histogram data appearing in black corresponds to the pre-fire image, while the gray one corresponds to the post-fire image. It is evident that among spectral channels of Landsat TM, TM4 and TM7 offer the highest discriminator ability to distinguish the burnt surfaces from the other land cover/use categories.

satellite data or a single post-fire image; second, considering if direct estimation of the burnt land is pursued or the techniques are applied just as an enhancement technique; and third, according to what type of training, supervised or unsupervised, is performed.

The first group of techniques is based on whether multitemporal or single post-fire satellite data are used. If a multitemporal data set is utilised, then the method is used in the context of the change detection approaches. On the other hand, if only a single post-fire satellite image is utilised, then the method relies mainly on the spectral behaviour of the burnt areas compared to the other land cover/use categories present in the study area. It has been demonstrated that the methods which utilise a multitemporal data set are more effective than those that utilise only a single post-fire image, since the former minimise the confusion between some permanent land cover types that present a similar spectral behaviour (Pereira et al. 1997). However, single post-fire methods present a superior performance over the multitemporal ones in the context of cost and time requirements needed for the acquisition and processing of the multitemporal data set. One of the most critical issues in the multitemporal approach concerns the radiometric and geometric adjustments which should be done in order to assure the spatial and spectral matching of the images used. Misregistration of both the radiometric and the geometric dimension, may produce unpredictable errors which in turn may result either under or over-estimation of the burnt areas. This of course depends on other parameters such as the size, patchiness, etc. of the burnt areas.

The second group of techniques is based on whether the output of the technique is the mapping of the burnt areas or a spectrally enhanced intermediate data set which needs further, although simple, processing. In remote sensing applications, especially those that utilise a multidimensional character data set, such as in the case of Landsat-5 TM, multivariate statistical methods are widely applied to extract the desired information. These methods, especially those dealing with the reduction of dimensionality, such as principal component analysis, vegetation indices, etc., aim to separate the spectral information distributed in the original spectral channels into a few new components which are more interpretable. If the reduction of the dimensionality and separation of the information is accomplished successfully, then by applying further simple processing, such as thresholding, the desired information can be easily achieved (Koutsias et al. 1998).

Finally, the third group of techniques is based on whether the method employs a supervised procedure, such as a maximum likelihood classifier, or a semi- or fully automated classification. Several methods which utilise both multitemporal or single post-fire satellite data require a thorough knowledge of the spectral properties of the objects. Many of the techniques employed in burnt land mapping require a careful and detailed definition of the training areas on the satellite images. The extracted spectral signatures from these training areas are then used to train the classifier or alternatively to build a model.

9.4.2
Description of the techniques

9.4.2.1
Principal component analysis

Theoretical aspects. Principal component analysis (PCA), a well-known dimensionality reduction technique, has been extensively applied in remote sensing studies. It produces a new uncorrelated data set, where the first component contains most of the original data variance, while the succeeding ones contain decreasing proportions of data variation (Richards 1984). PCA is an appropriate technique to explain the variance-covariance structure of an initial set of variables by the construction of new components, which are linear composites of the original ones. The primary goal of the analysis is the reduction of the dimensionality of a data set and the removal of the correlation among the variables (Sharma 1996, Johnson and Wichern 1998). Both of these issues are further associated with improvement in data interpretability and the enhancement of some particular structures distributed in the original data set. A fundamental issue involved in PCA is the presence of correlation among the initial variables, such as those of Thematic Mapper, which reflects the repetition of the information. The principal component aims to produce a new data set, through a linear algebraic expression of the initial variables, to minimise the correlation and associate the variance of the data with the new first components. Under this aspect, PCA is applied in change detection analysis, to produce a new enhanced data set, where the variance associated with permanent landscape features is accumulated in the higher order components, while that of the changed features is emphasised in the lower ones (Fig. 9.3). For the computation of the principal component three steps are involved. The first is the derivation of the covariance matrix (non-standardised PCA) or the correlation matrix (standardised PCA), the second is the estimation of the eigenvectors and the third is the linear transformation of the original data set

(Richards 1986). The values of the covariance or the correlation matrix can be derived either from the total study area or from a subset. In this sense, PCA is considered to be a scene-dependent technique, which requires very careful appraisal (Fung and LeDrew 1987).

Case studies. In burnt land mapping, PCA has been applied in the context of change detection using a multitemporal data set consisting usually of pre- and post-fire satellite images (Richards 1984, Milne 1986, Pereira 1992, Martin et al. 1994). There are also applications of single post-fire images (Tanaka et al. 1983) or a selected subsequence of the spectral channels (Richards 1984, Pereira 1992, Siljestrom and Moreno 1995). Since the burnt surface is considered to be a non-stable area when referred to the multitemporal context, it is expected to be found on the secondary components deducted from PCA, usually on the third or the fourth (Fig 9.3).

Concluding remarks. The reduction of the dimensionality achieved by the principal component transformation has been proved very useful in delineating the spectral information of the areas that have changed in time, such as those affected by wildland fires (Siljestrom and Moreno 1995). As principal component analysis is considered to be a scene-dependent technique, its performance, especially when used in the framework of change detection, is influenced by some factors which characterize the local scene. Richards (1984) stated that the sub-scene, which includes the changed region, should also include a substantial region of relatively no change, to ensure that the first components are associated with the variance found on stationary cover types.

9.4.2.2
Spectral mixture analysis

Theoretical aspects. The reflected energy recorded by the satellite sensors operating in certain spatial resolutions is the result of the interactions between the incident solar radiation and the various sub-objects found at the pixel level, influenced also by some other factors such as the presence of atmosphere, etc. As a result, the reflectance values of the pixels are composed of a mixture of various sub-homogeneous reflectances rather than of a totally pure one. Even when the earth surface is captured at finer spatial resolutions, the reflected energy is still a mixture of smaller components with differing spectral behaviour although the mixture is not so extended.

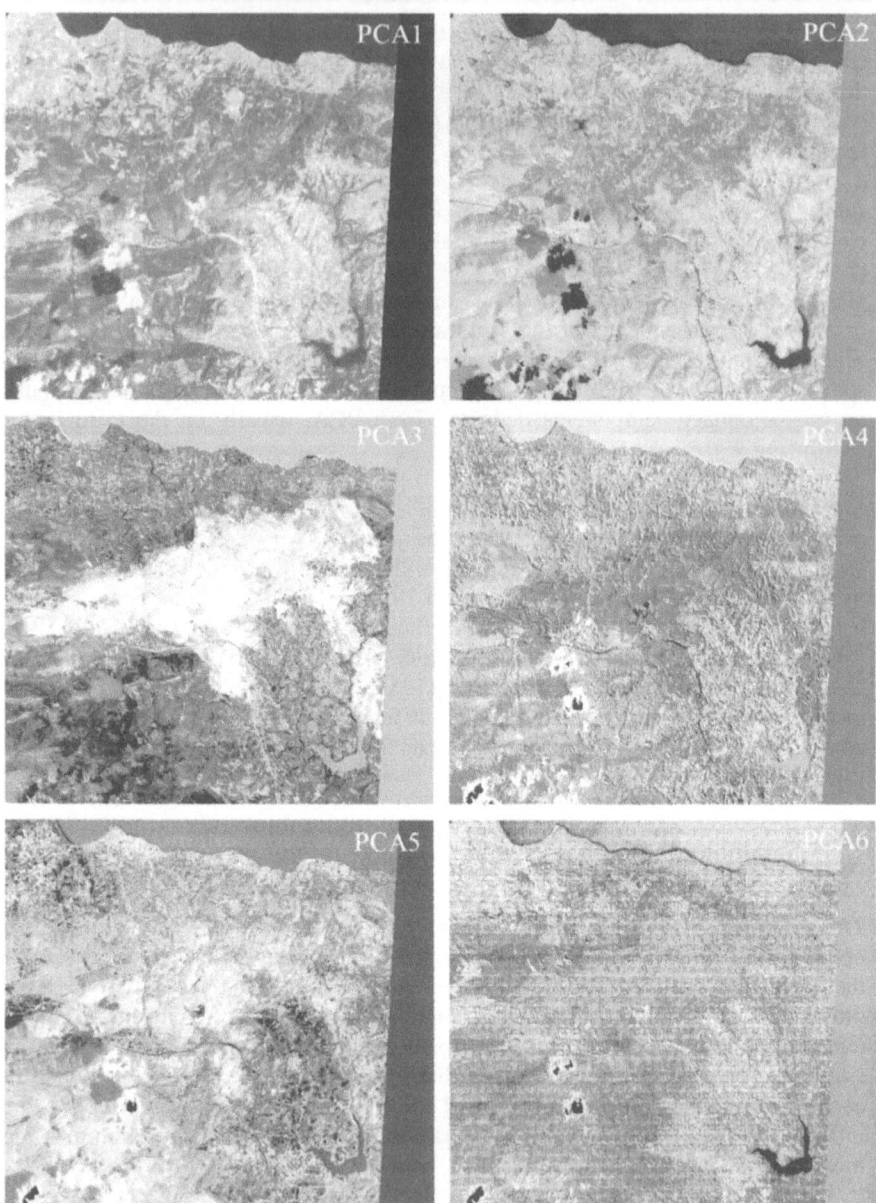

Fig. 9.3. Principal components of the post-fire satellite image (Plate 9.3). The variance associated with permanent landscape features, which occupy a large sub-scene of the satellite image, is accumulated in the first and second components, while that of changed features, such as the burnt surfaces, is emphasised in the third component. The last components consist mainly of noise.

Under this aspect, spectral mixture analysis (also known as spectral unmixing) has been developed and used in remote sensing studies to address the problem of mixed pixels. Its basic assumption is that the spectral variability found on the satellite image at the pixel level is the result of the mixture of smaller components with differing spectral behaviour (Adams et al. 1986, Caetano et al. 1994). Spectral mixture analysis, actually, aims to extract pure spectral signatures from such complex structures occurring at the pixel level and express them in concentrations of reference endmembers (Ustin et al. 1993, Pereira et al. 1997). In particular, the output of the analysis is a fraction image for each endmember for which the error associated with the estimations is also provided. The implementation of the method necessitates three steps; model building, model fitting and fraction image editing. Model building and fitting, which are two interrelated issues, include the assessment of the endmembers and the way in which they interact to estimate the pixel fractions (Caetano et al. 1994).

Case studies. Spectral mixture analysis has been applied quite successfully for burnt land studies using Landsat TM imagery (Caetano et al. 1994) and NOAA-AVHRR data (Caetano et al. 1996). In the former, a spectral mixing model was developed for both burnt land and fire severity mapping, which converts the radiance of the TM channels into proportions of soil, unburnt and burnt vegetation at a sub-pixel level. The study area, which was located in a mountainous area in central Portugal, gave the chance to explore the potentiality of the technique and to address some common problems found in Mediterranean-type landscapes (Caetano et al. 1994).

Concluding remarks. According to Caetano et al. (1994), spectral mixture analysis can contribute to solving some problems occurring when techniques such as statistical classification, band ratioing, etc. are involved in burnt land mapping. Among them we can distinguish reduction in topographic effects, since shade can be considered as an endmember, and the overcoming of the mixed pixel problems.

9.4.2.3
Logistic regression modelling

Theoretical aspects. To predict a dependent response variable from a set of independent explanatory measurements or to classify individuals into a group of two or more categories, various multivariate methods can be applied, although they presuppose the validity of a certain set of assumptions, among which the assumption of univariate and multivariate normality is the most frequent. It is violated when the dependent or the independent measurements is a mixture of categorical and continuous variables. Logistic regression, a method developed and used in survival analysis where the dependent response variable is binary dichotomous, can be applied as an alternative classification method when neither the multivariate normality is assumed nor the independent measurements consist only of continuous scalar variables (Afifi and Clark 1990, Norusis 1990, Mendenhall and Sincich 1996). The logistic model, which associates a dependent response variable with a set of independent explanatory variables, may be adapted for burnt land mapping using remote sensing data following (Mendenhall and Sincich 1996):

$$E(y) = \frac{\exp(b_0 + b_1X_1 + b_2X_2 + \ldots + b_kX_k)}{1 + \exp(b_0 + b_1X_1 + b_2X_2 + \ldots + b_kX_k)}$$ (1)

or equivalent (Afifi and Clark 1990):

$$E(y) = \frac{1}{1 + \exp[-(b_0 + b_1X_1 + b_2X_2 + \ldots + b_kX_k)]} ,$$ (2)

where:

$y = 1$ if the pixel is burnt or $y = 0$ if the pixel is not burnt

$E(y) = P(\text{pixel is burnt}) = \bullet$

$X_1, X_2, \ldots, X_k$: radiometric reflectances or any other multispectral transformation of the 1st, 2nd, …, kth spectral channel

$b_1, b_2, \ldots, b_k$: estimated coefficients of the model

The curve which results from the logistic regression equation has two asymptotes; a minimum at the value 0 and a maximum at the value 1 (Fig. 9.4). Thus the probability estimates always range between 0 and 1, which forms a realistic probability surface (Narumalani et al. 1997).

Actually, the intermediate values, in respect to burnt land mapping, reflect the probability based on which a candidate pixel belongs to the burnt category. The nature of the topic of burnt land mapping fits with the particular objectives of the logistic regression modelling, since the dependent variable can be expressed in a dichotomous way, that is, a candidate pixel to be burnt or unburnt. For this, a dummy or indicator variable E is created which takes the value 1 if the respondent is burnt or 0 if the respondent is unburnt. The independent explanatory measurements can be either the radiometric values of the original spectral channels or they can be transformed values arising from multispectral transformations such as vegetation indices, principal components, etc. The criterion used to classify an individual candidate pixel into one of the two mentioned dependent observations depends on the value of E(y). In a general form, when E(y) is larger than 0.5, then the individual pixel is classified as burnt, otherwise as unburnt; however, intermediate values of E(y) may indicate discrete levels of burning degree (Fig. 9.5). The latter is useful when a more detailed mapping of the burnt land is assumed, such as in the case of mapping the levels of vegetation damage.

To assure the successful development of the logistic regression model a certain set of prerequisites for the sampling process should be provided (Koutsias and Karteris 1998a, 1998b). These are the following:

- Accurate location of the sampling areas especially those of burnt cases on the satellite images used.
- Similar sampling size of the burnt and unburnt observations, to avoid bias in the estimation of the model coefficients.

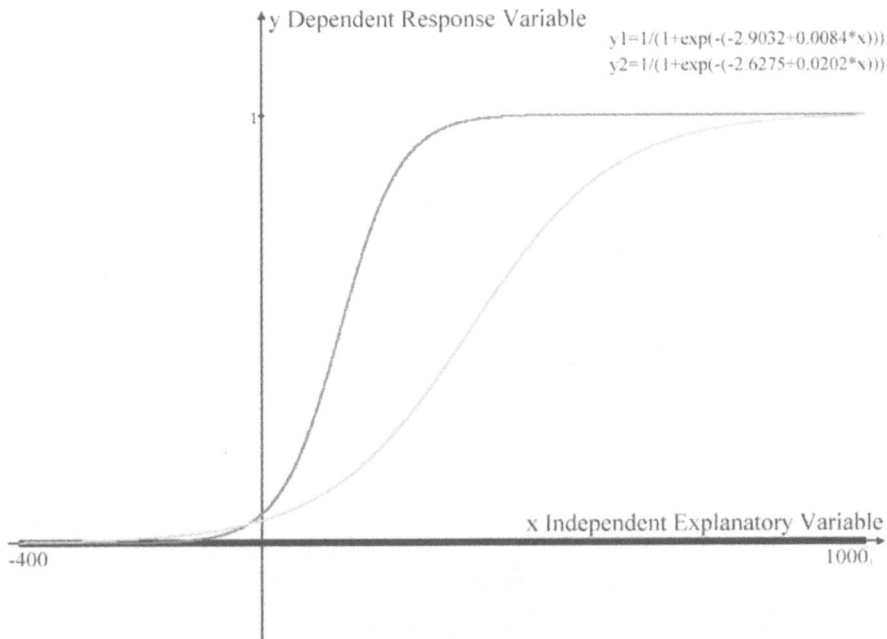

Fig. 9.4. The curve which results from the logistic regression equation has two asymptotes; a minimum at the value 0 and a maximum at the value 1. Thus, the probability estimates always range between 0 and 1, which forms a realistic probability surface.

• A satisfactory absolute sampling size for both cases, to assure the representative sampling and to cover all the variability found on the satellite images within and outside the burnt areas.

Case studies. Logistic regression modelling has been successfully applied in burnt land mapping using a multitemporal data set (Koutsias and Karteris 1998a), as well as using a single post-fire image (Koutsias and Karteris 1998b), both acquired from Landsat-5 TM (Plate 9.3). The overall maximum accuracy achieved after the application of the models was 97.62% and 97.37%, respectively, indicating the high performance of the methods. It should be noticed that the post-fire satellite image used was acquired a few days after the fire. Moreover, these two studies explored the usefulness of logistic regression modelling in assessing the overall discriminator ability offered by each spectral channel of Landsat-5 TM.

Concluding remarks. Logistic regression modelling, applied as an alternative classification technique, proved to be very useful in burnt land mapping, since the classification target could be expressed in a dichotomous way, that is as a candidate pixel to be burnt or unburnt. These studies demonstrated also the usefulness of the technique in assessing the spectral information content and the potential discriminator ability offered by each spectral channel of Thematic Mapper in burnt area mapping.

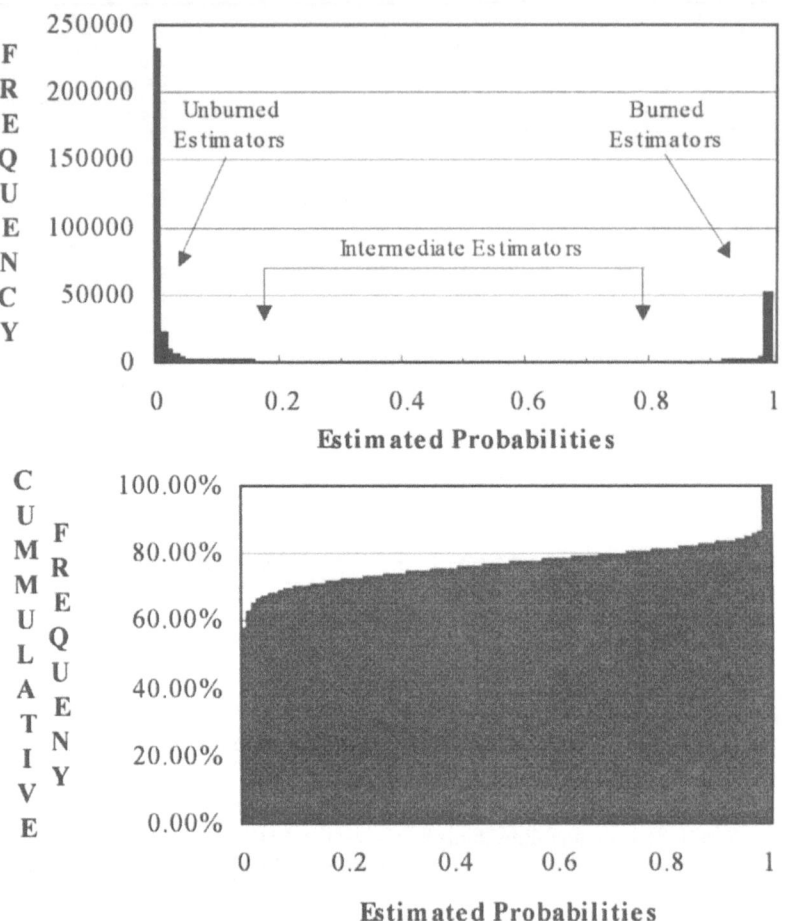

Fig. 9.5. The intermediate estimators, in respect to burnt land mapping, reflect the probability based on which a candidate pixel belongs to the burnt category. Moreover, these intermediate values may indicate discrete levels of burning degree.

9.4.2.4
Intensity-Hue-Saturation transformation

Theoretical aspects. Red-Green-Blue (RGB) and Intensity-Hue-Saturation (IHS) are two colour models that are widely applied to represent the colour on electronic or other devices. Intensity refers to the total brightness of the colour, hue refers to the dominant wavelength of the light and saturation refers to the purity of the colour (Mather 1987, Carper et al. 1990, Lillesand and Kiefer, 1994). The coordinates of the RGB model range between 0 and 1 for each axis, while for the IHS transformation the values range between 0 and 1 for intensity and saturation component and between 0 and 360 for hue component (Edwards and Davis 1994). To transform the values of an RGB colour composite to the IHS components, several different algorithms have been developed, which differ in their way of calculating the intensity component and in the RGB colour used as reference point for calculating the hue component. In general, the IHS transforms the values of a three channel composite in such a way that the spatial information is separated in the intensity component, while the spectral information is separated in the hue and the saturation components (Carper et al. 1990).

Case studies. Intensity-Hue-Saturation (IHS) transformation, a method mainly used for mapping multiresolution and multispectral data and for contrast-stretching applications, has been applied for burnt land mapping using a three-channel composite of Landsat-5 TM (Koutsias et al. 1998). The hue component of two RGB colour composites consisting of the Thematic Mapper channels TM7, TM4 and TM1, proved to be very useful in burnt land mapping since the fire scar, compared to other land cover categories, was well differentiated (Fig. 9.6). The reduction in dimensionality of the original spectral space and the elimination of the information responsible for discrimination of the burnt surfaces was accomplished successfully, so that two very distinct groups of pixels, one belonging to the burnt category and the other to the unburnt, appeared in the hue component. In this study, the algorithm to transform the RGB to IHS values, developed by Conrac (1980) and described in the Erdas field guide, was adopted (for further details see Erdas Inc. 1991).

Two among many RGB colour composites that were transformed to IHS values proved to be very useful in burnt land mapping. They were the composites of TM7-TM4-TM1 and TM4-TM7-TM1 of Landsat-5 TM. Actually, in the hue component the fire scar was well discriminated from other land cover/use categories presented in the satellite image (Fig. 9.6).

Koutsias et al. (1998) pointed out some reasons in the attempt to explain why the burnt surfaces were well discriminated in the hue component and not in the intensity or saturation. These reasons mainly depend on the mathematical expressions involved in the transformation, on how each component utilises the initial spectral information and on what each component of the transformation actually represents. As already mentioned, intensity refers to the spatial information of the composite image, and hue and saturation to the spectral. Thus burnt areas are expected to be identified on the hue or saturation component, since they constitute a spectral, rather than a spatial pattern. On the other hand, the values of the hue component of the RGB colour composites used result from the spectral information of TM7 and TM4, which has been proved the most suitable for burnt area mapping (Lopez and Caselles 1991, Koutsias and Karteris 1998a, 1998b).

So far, the IHS transformation has been mainly applied for merging satellite data acquired from different sources, by applying a forward/inverse transformation in which the intensity component is replaced by the higher spatial resolution image (Blom and Daily 1982, Haydn et al. 1982, Thormodsgard and Feuquay 1987, Welch and Ehlers 1987, Carper et al. 1990, Chavez et al. 1991), and as an image enhancement technique (Haydn et al. 1982, Green 1983, Gillespie et al. 1986, Edward and Davis 1994) especially in the context of geological applications (Mather 1987).

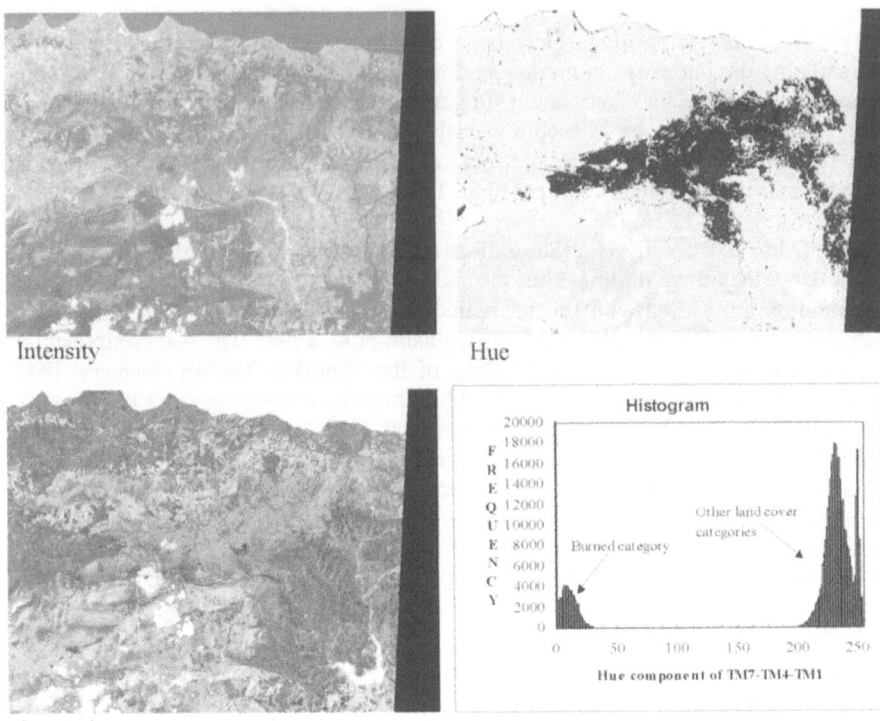

Saturation

Fig. 9.6. The Intensity, Hue and Saturation component of the IHS color model of the TM7-TM4-TM1 RGB color composite. In the hue component the burnt areas are well discriminated from the majority of the land cover/use categories. Confusion with some of the categories still remains as a problem, although it is not so extended. Among the confusing categories we can distinguish some areas dominated by sparse vegetation, the coastal line, and some segments of the road which passes through the forested area.

Concluding remarks. The Intensity-Hue-Saturation transformation, a method mainly used for merging multiresolution and multispectral satellite data and for contrast-stretching applications, has been successfully applied for burnt land mapping. The reduction of the dimensionality and the elimination of the spectral information was accomplished successfully. Actually two very distinct groups of pixels, one belonging to the burnt category and the other to the unburnt, appeared in the hue component of the TM741 RGB colour composite.

The IHS transformation proved to be superior to other methods in the following aspects (Koutsias et al. 1998):

- It does not require radiometric corrections or radiometric enhancements
- It does not require definition of training areas
- It produces a new data set in which the burnt areas are well discriminated
- Confusion between burnt areas and other land cover/use categories, such as shadows, urban areas, water bodies, etc. is severely reduced

9.4.2.5
Other techniques

Besides the implementation of the techniques described previously, other methods employed in burnt land studies include visual analysis and estimation, incorporation of vegetation indices, application of supervised and unsupervised classification procedures, involvement of linear regression, etc.

Among these, visual analysis is the simplest and under some circumstances can prove very effective for burnt land mapping, especially when used for masking the broad area around the fire extent before digital processing of the satellite data (Chuvieco and Congalton 1988). In this study, the RGB colour composite used consisted of the middle and near infrared spectral channels of Landsat-5 TM (4, 7 and 5). However, in other studies a superior performance of the TM 7,4,1 colour composite was also noticed (Koutsias and Karteris 1998b). The opportunity to interpret, apart from the spectral resolution, the spatial arrangement of the objects in the satellite image is considered to be its main advantage. As already mentioned, spectral similarities between the burnt surfaces and some other land cover/use categories occur very commonly, which result in confusion and misclassification among them. An effective solution to overcome this spectral confusion is to take into consideration the textural and contextual information (Pereira et al. 1997), which well-trained personnel can easily elaborate. However, this exhibits some disadvantages in respect to time requirements and the subjectivity of the interpretation, which limit its efficiency.

Vegetation indices, especially when used in a multitemporal approach, have been widely applied in burnt land mapping, either integrated with other channel combinations into more elaborated classification procedures (Pereira et al. 1997) or alone using change detection techniques (Milne 1986, Chuvieco and Congalton 1988, López and Caselles 1991, Kasischke et al. 1993, Viedma et al. 1997). Vegetation indices, which result from the algebraic combinations of the spectral channels, reflect the vegetation vigour and attempt to estimate several vegetation parameters (Tucker 1979, Campbell 1987). The aim is to reduce the multispectral observations to a single numerical index, which has been found to be well correlated with the internal factors of the ecosystems components, such as biomass, leaf area index, vegetation cover, etc. and also to minimise the effect of external factors such as inclination, sun orientation, etc. (Jensen 1986, Buret and Guyot 1991, Wiegand et al. 1991). Among them, the Normalised Difference Vegetation Index (NDVI) has been proved to be very useful for burnt land mapping (Fig. 9.7), although in other studies a different version of this index resulting by the replacement of TM3 with TM7 (Fig. 9.7) showed superior performance (Lopez and Caselles 1991, Koutsias and Karteris 1998b).

Supervised or unsupervised classification is a very common and well-explored method which has been applied for a wide range of applications. In the framework of burnt land mapping, the first attempts employed a digital supervised classification of a pre and post-fire satellite image followed by a comparative evaluation of the classification results. Chuvieco and Congalton (1988), when applying a traditional supervised classification in a Thematic Mapper image to study a large forest fire in Spain, reported some significant inaccuracies in the classification results. However, refining the selection of the training statistics

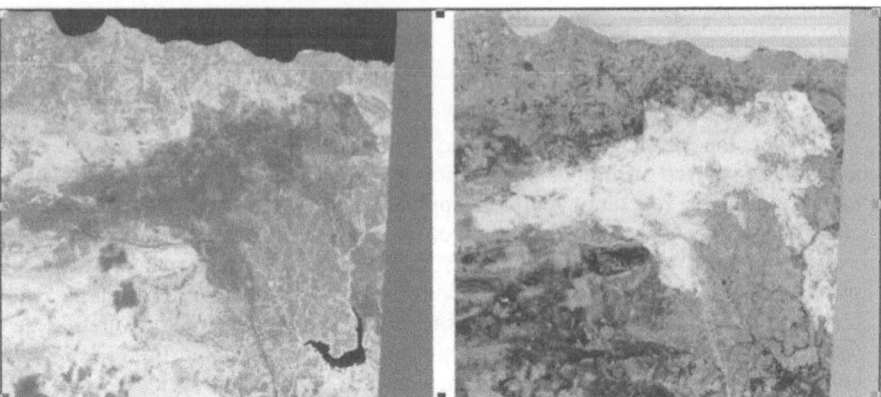

Fig. 9.7. The NDVI (left image) of the post-fire satellite image, and a modified version of the NDVI which results by replacing the TM3 with TM7 (right image). The superior performance of the latter index over the NDVI to enhance the spectral discrimination of the burned surfaces is evident.

using a supervised-unsupervised technique and also integrating certain channel ratios with the original spectral bands, brought a notable improvement in the classification performance.

9.5
Discrimination of damage intensities

9.5.1
Interest in discriminating damage intensities

As previously described, different image processing techniques make it possible to discriminate fire perimeter by the use of high-resolution imagery. This chapter deals with classification of levels of fire damage within the scorched area, which may be a difficult task because of the many factors that play an important role in the signal received by the sensor. Among these, the most important are:

- the complex spatial patterns of the mediterranean vegetation, where different strata and vegetation communities are usually mixed,
- the degree of carbonisation,
- the type of damage (trunk damage, surface damage, crown damage),
- proportion of ashes and soil,
- proportion of damaged and non damaged vegetation within pixel.

In fact, due to the different situations resulting from the above mentioned factors, classification of a burnt area in terms of damage levels even by fieldwork may become a difficult task, either quantitatively (carbonisation degree) or qualitatively (general categories of damage).

High resolution satellite imagery displays a synthetic view of these complex situations, making possible a quicker and cheaper evaluation of the affected area than the one provided by traditional methods of fire effects assessment (field inventories). This advantage can be highly appreciated in a forest management context, where simple maps which can be quickly supplied are required for the decision-making process.

Basically the techniques used in the classification of damage do not differ from those applied to burnt land mapping, either by the use of single post-fire images or by multitemporal information. Anyway, in most of the cases, validation of the satellite imagery with fieldwork and/or aerial photography was estimated as necessary. As the most common methods the following can be mentioned:

Thresholding either of single bands (Hall et al. 1980) or different Vegetation Indices (Dagorne et al. 1990, Navarro 1991, Salas et al. 1997, Rodríguez y Silva et al. 1997)

- Multivariate analysis of original bands (Tanaka et al. 1983)
- Unsupervised classification (Benson and Briggs 1978, Lobo et al.1998)
- Supervised classification of original bands, and or Vegetation Indices (Milne 1986, Jakubauskas et al. 1990) and ancillary information (Navarro et al. 1997b).
- Supervised-unsupervised classification of NDVI images (Chuvieco 1989, Chuvieco and Congalton 1988).

Within the Megafires Project, taking into account the frame of an operational and management-focused system, different processing techniques were applied in Andalusia (Spain) for the 1995 and 1997 fire seasons .

Previously, a visual Damage Index for field survey was specifically designed. This Damage Index, based on visual qualitative data, distinguished five categories of damage:

- no damage: no damage can be observed on the vegetation,
- low damage: surface damage, with trees either slightly affected or not affected and the understory slightly affected,
- moderate damage: understory and trees have been affected to a great extent but most of the vegetation has not been destroyed, some green crowns can be seen in between,
- high damage: trees have been destroyed, some foliage still remains in the tree crowns,
- extreme damage: vegetation is totally scorched, crowns totally destroyed.

The Damage Index allowed easy classification of vegetation in the field in sample areas, which would be used to test the accuracy of the different techniques utilised.

9.5.2
Description of the techniques

9.5.2.1
Vegetation Indices thresholding

Theoretical aspects. Based on the proven relationship between Vegetation Indices and plant parameters, application of different Vegetation Indices to discriminate damage levels was carried out as the first step for the 1995 fire season.

Vegetation Indices thresholding for fire damage discrimination is based on the assumption that there is a linear relationship between the different ranges of Vegetation Indices and a level of damage. For this reason, prior to performing this technique three Vegetation Indices (NDVI, SAVI, ARVI) were tested versus field data as a way to test determine to what extent each damage level would be related to the values of Vegetation Indices (VI).

SAVI and NDVI are VI commonly used in vegetation studies, but ARVI is not so common. This index was proposed by Kaufman and Tanré (1992) to reduce the atmospheric influence on the vegetation signal, by considering radiance in the blue band, which is most affected by atmospheric scattering. ARVI is computed from the following expression:

$$ARVI = (Lnir - Lrb)/(Lnir + Lrb)$$
$$Lrb = Lr - f(Lb - Lr),$$
(3)

where *Lnir, Lr, Lb* are radiances in the near infrared, red and blue TM bands, respectively; *f*, type of aerosol (a value of 1 can be taken if no data are available; also for the study values of 0.5 and 2 were taken).

From the outcoming results it was concluded that the different Vegetation Indices could be grouped into two categories according to VI sensitivity to describe damage level: NDVI and SAVI in the first place, and ARVI (with 0.5, 1 and 2 values for *f*) in the second. NDVI and SAVI had similar accuracy values, with 57 and 56%, respectively. ARVI reached very low values of accuracy, thus indicating a very low sensitivity to detect fire damage.

Only for totally burnt and non-burnt areas (extreme damage and no damage) was it possible to find a straight 'universal' relationship between VI values and the Damage Index, which was not possible for middle categories (moderate and high damage), whose spectral values fluctuated in the different points. This was partly due to the different proportions of the damaged and non-damaged trees within the pixel and partly to the definition of these categories in the Damage Index. Reduction of the initial five categories of the Damage Index to 3 was pointed out as a way to improve the values of accuracy of the three Vegetation Indices.

Case studies. From the first results of the 1995 fire season, thresholding of the NDVI and modified NDVI (with Band 5 instead of Band 3) was attempted for a 1997 fire season forest fire in the Los Barrios area (Andalucía, Spain).

Results and concluding remarks. Figure 9.8 shows the mean, standard deviation and standard error values in the NDVI and modified NDVI in relation to damage values for five and three categories of damage. With respect to NDVI, the index shows a decreasing tendency as the damage degree increases, though this decreasing tendency seems to change radically in the 4 and 5 damage levels. On the other hand, the modified NDVI shows an opposite tendency and it seems to fit an exponential function. Although some meaningful differences can be described between both vegetation indices, an internal overlap can be remarked, which makes it difficult to establish the limits between the different damage levels. Simplification of original damage levels into three classes decreases this internal overlap, as can be seen in Fig. 9.8. In this case thresholding of the NDVI and the modified NDVI in three ranges shows a 45% and 42% level of accuracy.

Although thresholding of VI provides higher levels of accuracy when the original five classes are reduced to three, they still remain very low, especially if they are compared with other methods of image processing. For this reason this technique should only be used under special conditions, when no field data are available and no very high level of accuracy is required. For most cases other image processing techniques should be applied.

9.5.2.2
Unsupervised classification: segment-based classification

Theoretical aspects. Assessment of fire impact can also be made by comparison of both pre- and post-fire images, which may entail some difficulties. One of them is that post- and pre-fire images often are separated by a significant time interval due to acquisition, atmospheric and/or economic reasons. Such a time lag involves some surface change on reflectance that is not due to fire and which is also dependent on the vegetation type. This is particularly important for Mediterranean vegetation, that is typically water-stressed. Consequently, images acquired a few weeks apart might show significant differences in vegetation reflectance if rain has occurred in between. Therefore, differential responses to fire and environmental conditions, along with the subtlety of the variable of interest, advise the use of detailed pre-fire vegetation maps for an stratified analysis. On the other

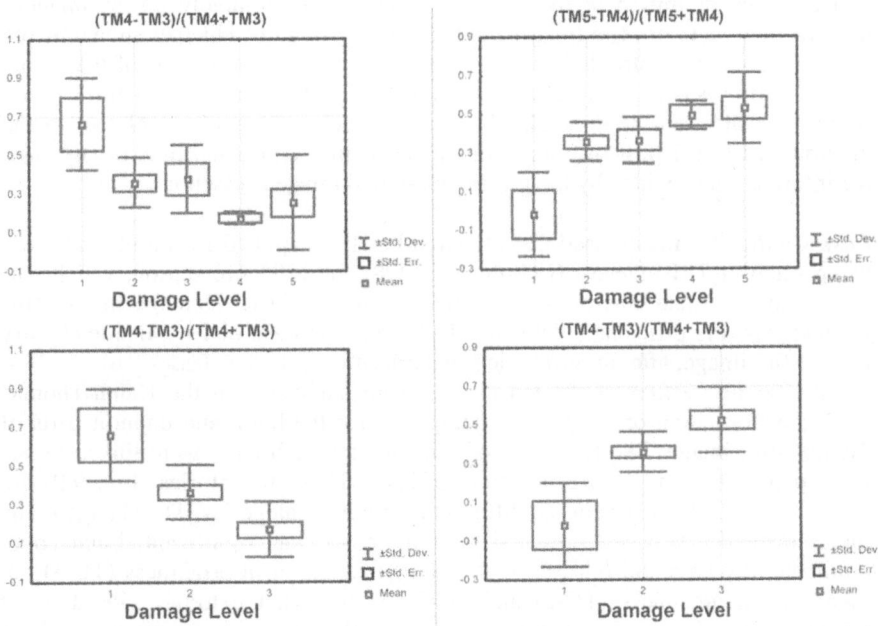

Fig. 9.8. (above) Mean, standard deviation, and standard error in NDVI and modified NDVI for five damage levels; and (below) Mean, standard deviation and standard error in NDVI and modified NDVI for three damage levels.

hand, the simulation of not-burnt conditions at post-fire time reflectance (Lobo et al. 1998) can greatly improve the assessment of fire damage intensity.

An exploratory analysis of the change between the pre- and the post-fire image is essential for an assessment of damages that could consider the eventual differential response of land cover categories. For this purpose it is necessary to define statistical strata by means of land cover classification. Methods based on image segmentation, hierarchical clustering and discriminant analysis have been found to be more adequate that conventional per-pixel classifications. This method is based on (i) a segmentation of the image (i.e. with the IMORM algorithm), (ii) a hierarchical clustering of a random subset of the image, (iii) a sub-optimal definition of the number of classes, (iv) a refinement of centroids (vectors averages and dispersion matrices) through an iterative linear canonical discriminant process, and finally (v) the classification of the entire image by allocating each segment to its closets centroid.

Once the pre-fire image has been classified, the centroids of each class can be calculated in both the pre- and the post-fire image and for both burnt and unburnt surfaces. A plot of the trajectories of the centroids from pre- to post-fire conditions in a reduced space is an important step for the interpretation of the spectral signatures and to define an ad-hoc index of fire impact. A common and useful reduced space is the one provided by the first three components of the Kauth-Thomas transform, but dedicated transforms based on canonical analysis are also useful.

The actual definition of the damage intensity from remotely sensed imagery must be relative to pre-fire images but take into account the changes shown by the trajectories of the centroids between both dates on the unburnt part of the scene. An interesting way of doing so is to simulate the post-fire image as if fire had not occurred, which can be done simply by using the medians of the classes in the unburnt part of the post-fire image or by means of a regression tree. In any case, the information provided by the pre-fire classified image is essential.

Case studies. Segment-based classification has been applied for burnt land areas using Landsat TM imagery (Lobo et al. 1998) for 1995 fire season, which was tested with the data classified in the field with the Visual Damage Index. This method was very efficient at detecting the fire scar and was also applied to classify the pre-fire image, after atmospheric standardisation, into ten classes.

The average values of the ten classes were calculated in the Kauth-Thomas (TC) transformation of the pre-fire image and for the burnt and unburnt parts of the post-fire image. The two pairs of averages for each class were linked by arrows pointing from pre-fire to post-fire dates. These trajectories (Fig. 9.9) are consistent with the nature of the different land cover categories. The change in the unburnt surface between pre- and post-fire dates was significant. Land cover dominated by bare soil (classes 1, 8 and 10) decreased in brightness (TC-1) but increased in greenness (TC-2) and TC-3. At the other extreme, closed forest (classes 4, 7 and 9) increased in brightness with decreased greenness and, slightly, TC-3. Open forest classes (2, 3 and 6) decreased in all three components. These dynamics imply that vegetation was drier at the post-fire date, while differences in bare soil reflectance are probably due to the change in solar incidence angle.

Arrows pointing from pre-fire to post-fire averages in the burnt surfaces describe change due to fire. Classes 1, 8, 10 and 5 (dominated by bare soil) did not get burnt, and actually acted as barriers to fire propagation. Closed canopy forest (classes 4, 7, and 9) severely decreased greenness and TC-3 with slight or no change in brightness, while open canopy forest (classes 2, 3 and 6) slightly decreased brightness in addition to a severe decrease in greenness. The decrease in brightness was more severe as the percent of openness was higher. These responses indicate that green vegetation lost the near-infrared - red contrast and that shortdry understory became darker.

This analysis indicated that the TC-2, the Kauth-Thomas greenness, could be used as a measure for the assessment of fire damage intensity. A regression tree was thus calculated to predict the unburnt TC-2 values of the actually burnt part of the post-fire image using the class, the TC-2 at pre-fire conditions and the terrain variables (height, slope and aspect) as predictors, and evaluated using a jack-knife procedure. Finally, the difference between the simulated and the observed TC-2 was used as an index of fire impact (Fig. 9.10).

Concluding remarks. The processing based on image segmentation is very efficient at detecting the boundary of the fire scar as well as at producing a land cover map of pre-fire conditions. Classes thus produced show a consistent response to phenologic, environmental and fire-induced change.

An estimation of fire impact cannot be done from a simple comparison of pre-fire and post-fire imagery if, as most often in Mediterranean landscapes, there is a significant time span between the dates of both images. The difference between the actual second Kauth-Thomas component and modelled values is significantly related to the field-estimated Damage Index, for which the discriminant power of the image and terrain variables already provide a cost-effective estimation of the degree of fire impact, in particular in the case that discrimination between the higher levels of damage is not a major concern for the end user.

9.5.2.3
Supervised Classification

Theoretical aspects. Related to this approach, the classification of intensities of damage has been made by the use of two different methods (Navarro et al. 1997b):
- Supervised minimum distance classification with the following combinations:
- Post-fire image with all bands.
- Post-fire image with all bands plus slope and aspect.

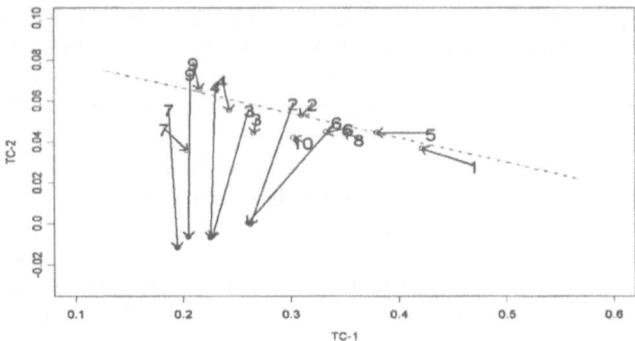

Fig. 9.9. Trajectories in the Kauth-Thomas space. Numbers represent median values for the 10 classes. Arrows point from values in the pre-fire TM image to the post-fire TM image. Open circles represent median values of the not-burnt part of the post-fire image, while solid circles represent the median values of the burnt part.

Post-fire TC-2 - Model Medians

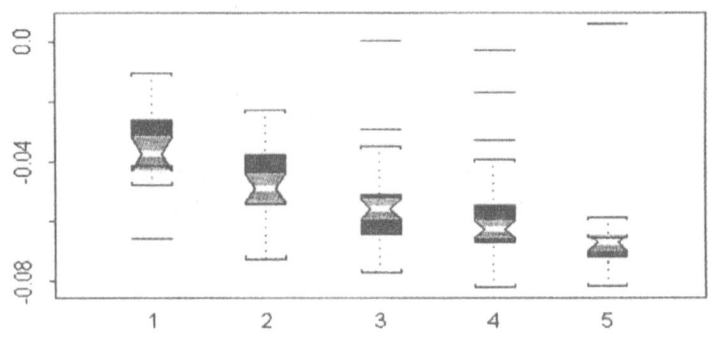

Level of Fire Impact (field)

Fig. 9.10. Box plots of the values of the index of fire impact (y-axes) stratified by the field-observed levels of fire impact (X-axes). White bands indicate the medians; square brackets, the range; notches, 95% confidence intervals of the medians.

- Post-fire and pre-fire images with all bands plus slope and aspect.
- Supervised maximum likelihood classification with a single combination of all TM bands of the post-fire image plus slope and aspect.

Supervised classifications are an application of the technique known as discriminant analysis. Through the analysis of the differences between two or more groups with respect to a set of variables simultaneously, each value will be classified in its closest group.

The basic conditions to perform this analysis are (i) the existence of two or more groups differing in some variables, (ii) a priori classification of the values has to be mutually excluding, (iii) each group has to be extracted from a population which is distributed according to a multivariate normal, and (iv) there is no variable that is a lineal combination of the others.

From the above mentioned conditions, discriminant analysis can be used to classify pixels of the scene. In this case a rule which optimally allows allocation of a new object to any of the established classes is needed. Basic statistics are: lineal canonical function, if the covariance matrices are identical, and the quadratic classification function, in the case that covariance matrices are not identical.

With the information gathered for each training area, supervised classification was performed on a post-fire image in the Los Barrios area (1997 fire season), with the original Landsat-TM bands (band 6 excluded), plus slope and aspect. Two classification rules were used: the first one was based on the lineal canonical function, which determines the values of these functions for each pixel (the number of functions is the same as the number of established groups). Finally, the pixel is classified in the group whose function has the highest value.

The second one is based on the maximum likelihood algorithm, which is expressed as follows:

$$D = \ln(a_c) - [0.5\ln(|Cov_c|)] - [0.5(X - M_c)^T (Cov_c^{-1})(X - M_c)],\qquad(4)$$

where D is the distance to class centre, c the particular class, X the vector with the pixel values, M_c: vector with the mean of class, a_c the a-priori probability of belonging to class c, and Cov_c the covariance matrix of class c.

Case studies. Both supervised classifications have been applied in two different fire areas of Andalucía (Spain) for 1995 and 1997 fire seasons: Aznalcollar (Sevilla) and Los Barrios (Cádiz). The results are mainly similar in both cases, as showed in Lobo et al. (1998).

Concluding remarks. Maximum likelihood classification of all original TM bands plus slope and aspect provides better accuracy values in relation to damage levels than other methods such as thresholding of different VI (NDVI, NDII) (Fig. 9.11).

The Landsat-TM bands which are more capable of discriminating between damage levels are (in decreasing order): 5, 7, 4 and 3. Regarding other variables which can be helpful, slope is more valuable than aspect. Therefore, for future work a combination of the 1,2,3,4,5 and 7 post-fire image bands plus slope is advisable.

The number of parcels to be surveyed for each fire is still under discussion. In any case the number of parcels per damage category should be enough to define the values of the co-variance matrix and the mean vector. Application of a maximum likelihood classification to a 7 bands image, at least 7+1 training pixels are required, aiming to avoid a singular co-variance matrix, which makes impossible the calculation of the discriminant function. According to Swain et al. (1978), a minimum of 10xN pixels should be taken, being N the number of bands to be used in the model.

Selection of suitable parcels in the field is essential for the classification stage, so that the following suggestions can be made:

- The land plot must be homogeneous in relation to the type of vegetation before the fire, damage level and illumination conditions.

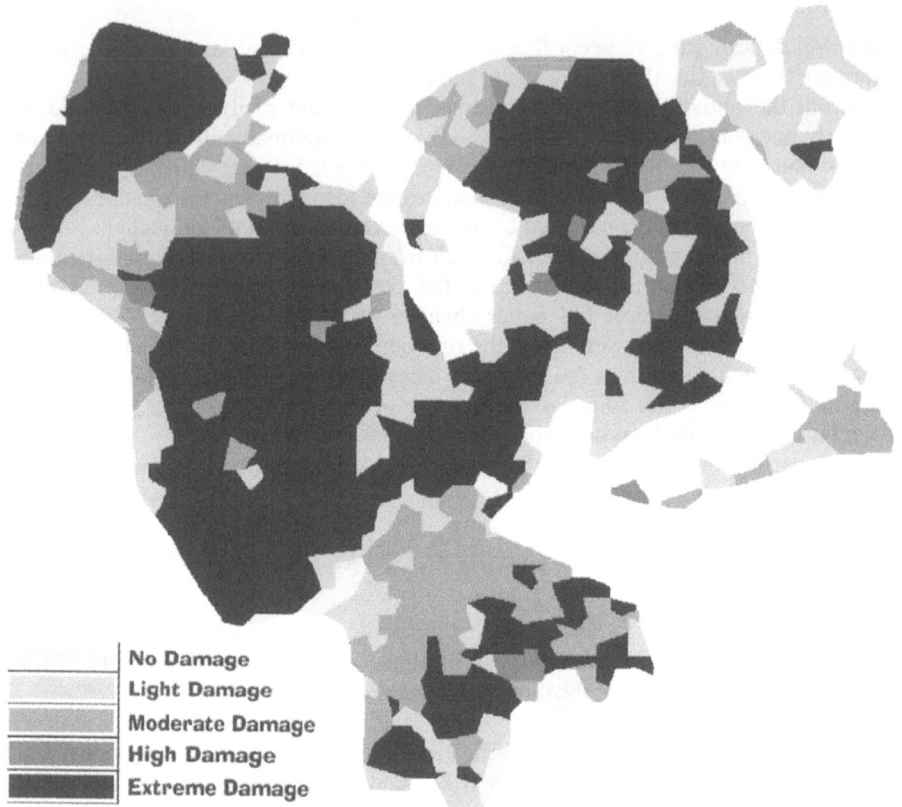

No Damage
Light Damage
Moderate Damage
High Damage
Extreme Damage

Fig. 9.11. Map of damage levels (Maximum likelihood classification).

- Plot size must be over 1 hectare so that a 3x3 pixel training area can be taken. Nevertheless, for a fire with a high mixture of pixels belonging to different damage levels, parcels should be slightly smaller.

As final conclusion, from an operational and management-focused point of view, some advantages of high resolution imagery in relation to discrimination of levels should be pointed out:

- Landsat pixel size (30 m) is very suitable to forest management and planning, as it is even smaller than the usual basic unit utilized in forest management in Andalusia, the rodal,
- the cost of Landsat images, together with the advantages of multitemporal information, makes them very useful for the evaluation of forest damage and regeneration monitoring at a reasonable price. Future sensors will, however, play an important role in forest fire evaluation in the coming years.
- Classification of pre and post-fire TM bands plus slope is the more useful technique, provided that field data are available.
- As addressed by other authors (Milne 1986), knowledge of vegetation conditions is essential to understand and interpret the nature of change caused by forest fires. Taking this into account, classification of scorched vegetation in

terms of regeneration potential is more suitable to the decision-making process than a damage level map. This can be easily approached with a simple integration of the vegetation and damage maps, so that different practices can be suggested for future management of the fire area.

9.6
Epilogue

Although forest fires may have a negative or a positive influence on the natural cycle of vegetation succession and on the ecosystem's structure and function, the tremendous number of forest fires frequently occurring over the same areas constitutes a real threat to natural ecosystems. The assessment of the consequences associated with fire activity, as well as the development of a well-structured decision-making system for the rational management of forest fires, imposes the development of a comprehensive information system including advanced and powerful monitoring processes.

Satellite remote sensing could effectively be involved in burnt land mapping, since it provides the necessary means of gathering information of the Earth's surface in a less expensive and timely fashion. Periodic spectral data in the visible and infrared part of the electromagnetic spectrum, of high spatial resolution, acquired from remotely satellite sensors, offer an unlimited basic source of information, which by appropriate, computer-assisted processing and interpretation can contribute to a better, cost-effective, objective and time-saving method to map and monitor areas affected by wildland fires.

Acknowledgements. The research reported in this chapter was developed under the Megafires research project (contract ENV4-CT96-0256), funded by DG XII of the European Commission.

References

Abhineet J, Ravan SA, Singh RK, Das KK, Roy PS (1996) Forest fire risk modelling using remote sensing and geographic information system. Current Science 70: 928-933

Adams JB, Smith MO, Johnson PE (1986) Spectral mixture modelling: a new analysis of rock and soil types at the Viking Lander I site. Journal of Geophysical Research 91: 8098-8112

Afifi AA, Clark V (1990) Computer-Aided Multivariate Analysis. Van Nostrand Reinhold Company, New York

Agee JK, Pickford SG (1985) Vegetation and fuel mapping of North Cascades National Park. Final Report. College of Forest Resources, Seattle

Aguado I, Chuvieco E, Camarasa A, Martín MP, Camia A (1998) Estimation of Meteorological Fire Danger Indices from Multitemporal Series of NOAA-AVHRR data, In: Viegas DX (ed) III International Conference on Forest Fire Research - 14th Conference on Fire and Forest Meteorology. ADAI, Luso-Coimbra, pp 1131-1147

Albini FA (1985) A Model for Fire Spread in Wildland Fuels by Radiation. Combust. Set and Tech 42: 229-259

Alcántara JM, Rey PJ, Valera F, Sánchez AM, Gutiérrez JE (1997) Habitat alteration and plant intra-specific competition for seed dispersers. An example with *Olea europaea* var. *sylvestris*. Oikos 79: 291-300

Alcázar J, Vega-García C, Grauet M, Pemán J, Fernández A (1998) Human risk and fire danger estimation through multicriteria evaluation methods for forest fire prevention in Barcelona, Spain, In: Viegas DX (ed) III International Conference on Forest Fire Research - 14th Conference on Fire and Forest Meteorology. ADAI, Luso-Coimbra, pp 2379-2387

Alexander, M.E. (1982) Calculating and interpreting forest fires intensities. Can J Bot 60:347-359

Alexander ME, McAlpine RE (1987) Canadian forest fire behaviour prediction (FBP) System field reference. Northern Forestry Centre. Western and northern region Canadian Forestry Service, Edmonton

Alonso M, Camarasa A, Chuvieco E, Cocero D, Kyun I, Martín MP, Salas FJ (1996) Estimating temporal dynamics of fuel moisture content of Mediterranean species from NOAA-AVHRR data. EARSEL Advances in Remote Sensing 4: 9-24

Amanatidis GT, Paliatsos AG, Repapis CC, Bartzis JG (1993) Decreasing precipitation trend in the Marathon area, Greece. Int J Clim 13: 191-201

Ambrosia VG, Brass JA (1988) Thermal analysis of wildfires and effects on global ecosystem cycling. Geocarto International 1: 29-39

Ambrosia VG, Buechel SW, Brass JA, Peterson JR, Davies RH, Kane RJ, Spain S (1998) An integration of remote sensing, GIS and information distribution for wildfire detection and management. Photogrammetric Engineering and Remote sensing 64: 977-985

Anderson Hal E.,1982. Aids to determining fuel models for estimating fire behavior. INT-122. USDA Forest Service, Ogden

Andrews PL (1986) BEHAVE: Fire behavior prediction and fuel modelling system. Burn subsystem. Part 1. USDA Forest Service, Ogden

Andrews PL, Chase CH (1990) The BEHAVE Fire Behavior Prediction System. The Compiler 8: 4-9

Arianoutsou M, Thanos CA (1996) Legumes in the fire-prone Mediterranean regions: an example from Greece. Int J Wildland Fire 6: 77-82

Arino O, Melinotte J-M, Calabresi G (1993) Fire, cloud, land, water: the 'Ionia' AVHRR CD-Browser of ESRIN. ESA, ESTEC, Noordwijk

Arnold K (1951) Uses of aerial photography in control of forest fire. Journal of forestry 49: 631

Aronoff S (1989) Geographic Information Systems: a management perspective. WDL Publications, Ottawa

Bachman A, Allgöwerm B (1998) Framework for wildfire risk analysis, In: Viegas DX (ed) III International Conference on Forest Fire Research - 14th Conference on Fire and Forest Meteorology. ADAI, Coimbra , pp 2177-2190

Baret F, Guyot G (1991) Potentials and limits of vegetation indices for LAI and APAR assessment. Remote Sensing of Environment 35: 161-173

Barredo JI (1996) Sistemas de Información Geográfica y Evaluación Multicriterio en la ordenación del territorio. RA-MA, Madrid

Belward AS (1991) Spectral characteristics of vegetation, soil and water in the visible, near-infrared and middle-infrared wavelengths. In: Belward AS, Valenzuela CR (eds) Remote sensing and Geographical Information Systems for resource management in developing countries. ECSC, EEC, EAEC, The Netherlands, pp 31-53

Belward AS, Grégoire J-M, D'Souza G, Trigg S, Hawkes M, Brustet J-M, Serça D, Tireford J-L, Charlot J-M, Vuattoux R (1993) In-situ, real-time fire detection using NOAA/AVHRR data. Proceedings of the 6th European AVHRR Data Users' Meeting, Belgirate, pp 333-339

Belward AS, Kennedy PJ, Grégoire J-M (1994) The limitations and potential of AVHRR GAC data for continental scale fire studies. International Journal of Remote Sensing 15: 2215-2234

Benediktsson JA, Swain PH, Ersoy OK (1990) Neural network approaches versus statistical methods in classification of multisource remote sensing data. IEEE Transactions on Geoscience and Remote Sensing 28: 540-552

Benson ML, Briggs I (1978) Mapping the extent and intensity of major forest fires in Australia using digital analysis of Landsat imagery. Proc on the Inter Symp on Remote sensing for observation and inventory of Earth resources, Freiburg, pp 1965-1980

Bian L (1997) Multiscale nature of spatial data in scaling up environmental models. In: Quattrochi PA, Goodchild MF (eds) Scale in remote sensing and GIS. CRC Lewis publishers, Boca Raton, pp 13-26

Bischof H, Schneider W, Pinz AJ (1992) Multispectral classification of Landsat images using neural networks. IEEE Transactions on Geoscience and Remote Sensing 30: 482-489

Blom RG, Daily M (1982) Radar image processing for rock-type discrimination. IEEE Transactions on Geoscience Electronics GE-20: 343-351

Blondel J, Aronson J (1995) Biodiversity and ecosystem function in the Mediterranean Basin: human and non-human determinants. In: Davis GW, Richardson DM (eds) Mediterranean-type Ecosystems: The function of biodiversity. Springer-Verlag, Berlin, pp 43-119

Bond WJ, van Wilgen BW (1996) Fire and Plants. Chapman & Hall, London

Botkin DB, Janak JF, Wallis JR (1972) Some ecological consequences of a computer model of forest growth. J Ecol 60: 849-872

Bourgeau-Chavez LL, Kasischke ES, French NHF (1994) Using ERS-1 SAR imagery to monitor variations in burn severity in an Alaskan fire-disturbed boreal forest ecosystem. Proceedings IGARSS'94, Pasadena, pp.243-245.

Bourgeau-Chavez LL, Harrel PA, Kasischke ES, French NHF (1995) The detection and interpretation of Alaskan fire-disturbed boreal forest ecosystems using ERS-1 SAR imagery. Proceedings IGARSS'95: Quantitative Remote Sensing for Science and Applications. IEEE Publications, Firenze, pp 1246-1248

Bovio G, Camia A (1997) Meteorological indices for large fires danger rating. In: Chuvieco E (ed) A review of remote sensing methods for the study of large wildland fires. Departamento de Geografía, Universidad de Alcalá, Alcalá de Henares, pp 73-89

Bovio G, Camia A (1998) An analysis of large forest fires danger conditions in Europe. In: Viegas DX (ed) III International Conference on Forest Fire Research - 14th Conference on Fire and Forest Meteorology. ADAI, Luso-Coimbra, pp 975-994

Bovio G, Quaglino A, Nosenzo A (1984) Individuazione di un indice di previsione per il pericolo di incendi boschivi. Monti e Boschi 35: 39-44.

Bovio G, Sol B, Viegas DX (1994) Synthese des travaux sur l'intercomparison des indices de danger meteorologique d'incendie. EC, Environment and Climate Programme, pp 22-38

Bowman WD (1989) The relationship between leaf water status, gas exchange, and spectral reflectance in cotton leaves. Remote Sensing of Environment 30: 249-255

Bradley AW, Yow CF, Heathcott M, Milne D, McCaffrey TM, Ghitter G, Franklin SE (1994) Landsat MSS classification of fire fuel types in Wood Buffalo National Park, Northern Canada. Global Ecology and Biogeography Letters 4: 33-39

Bradshaw L, Deeming J, Burgan RE, Cohen J (1983) The 1978 National Fire-Danger Rating System: Technical Documentation. GTR INT -169. USDA, Forest Service, Intermountain Forest and Range Experiment Station, Ogden

Brass J, Likens WC, Thornhill RR (1983) Wildland inventory and resource modeling for Douglas and Carson City counties, Nevada, using Landsat and digital terrain data. NASA, Scientific and Technical Information Branch, Washington, D.C.

Brotak EA (1980) A comparison of the meteorological conditions associated with major wildland fires in the USA, Canada and Australia. 6th Conference on Fire and Forest Meteorology, Seattle, pp 38-41

Brotak EA (1991) Low-level temperature, moisture and wind profiles preceding major wildland fires. Proceedings 11th Conference on Fire and Forest Meteorology, Missoula, pp 503-510

Brotak EA, Reifsnyder E (1977) Predicting major wildland fire occurrence, Fire Management Notes, pp 5-8

Brotak EA, Reifsnyder E (1976) Synoptic study of the meteorological conditions associated with extreme wildland fire behavior. Fourth National Conference on Fire and Forest Meteorology, St Louis, pp 66-69

Brown JK, Booth GD, Simmerman DG (1989) Seasonal change in live fuel moisture of understory plants in western U.S. Aspen. 10th Conference on Fire and Forest Meteorology, Ottawa, pp 406-412

Bugmann HKM (1997) Sensitivity of forests in the European Alps to future climatic change. Clim Res 8: 35-44

Burgan RE (1988) 1988 Revisions to the 1978 National Fire-Danger Rating System. Research Paper SE-273. USDA, Forest Service, Ogden

Burgan RE, Hartford RA (1993) Monitoring vegetation greenness with satellite data. USDA Forest Service, Ogden

Burgan RE, Klaver RW, Klaver JM (1998) Fuel models and fire potential from satellite and surface observations. International Journal of Wildland Fire 8: 159-170

Burgan RE, Rothermel RC (1984) BEHAVE: Fire behavior prediction and fuel modeling system. Fuel subsystem. USDA Forest Service, Ogden

Burgan RE, Shasby MB (1984) Mapping broad-area fire potential from digital fuel, terrain, and weather data. Journal of Forestry 82: 228-231

Byram GM (1959) Combustion of forest fuels. In: Davis KP (ed) Forest fire: control and use. McGraw-Hill, New York, pp 155-182

Caballero D, Martinez-Millan J, Martos J, Vignote S (1994) CARDIN 3.0. A model forest fire spread and fire fighting simulation. 2nd International Conference on Forest Fire Research, Coimbra, pp 501-502

Caetano OE, Mertes LAK, Pereira JMC (1994) Using spectral mixure analysis for fire severity mapping. 2nd Int Conf Forest fire research, Coimbra, pp 667-677

Caetano MS, Mertes LAK, Cadete L, Pereira JMC (1996) Assessment of AVHRR data for characterizing burned areas and post-fire vegetation recovery. EARSeL Advances in Remote Sensing 4: 124-134

Cahoon Jr, DR, Stocks BJ, Levine JS, Cofer III WR, Pierson JM (1994) Satellite analysis of the severe 1987 forest fires in northern China and southeastern Siberia. Journal of Geophysical Research 99: 18627-18638

Campbell JB (1987) Introduction to remote sensing. The Guilford Press, New York

Campbell J, Weinstein D, Finney M (1995) Integrating Landsat TM imagery, GIS, and BEHAVE for forest fire fuels mapping and fire behaviour modelling on the Camp Lejeune Marine Corps Base. ACSM-ASPRS Annual Convention, Charlotte, pp 365-373

Cao C, Lam N (1997) Understanding the scale and resolution effects in remote sensing and GIS. In: Quattrochi PA, Goodchild MF (eds) Scale in remote sensing and GIS. CRC Lewis publishers, Boca Raton, pp 57-72

Carper WJ, Lillesand TM, Kiefer RW (1990) The use of intensity-hue-saturation transformations for merging SPOT panchromatic and multispectral image data. Photogrammetric Engineering and Remote Sensing 56: 459-467

Carrega P (1990) Climatology and index of forest fire hazard in Mediterranean France. II International Conference on Forest Fire Research, Coimbra, pp B.05:1-11

Carreira JA, Arevalo JR, Niell FX (1996) Soil degradation and nutrient availability in fire-prone Mediterranean shrublands of southeastern Spain. Arid Soil Res Rehab 10: 53-64

Carter GA (1991) Primary and secondary effects of water content on the spectral reflectance of leaves. American Journal of Botany 78: 916-924

Castro R, Chuvieco E (1995) Clasificación digital de combustibles forestales a partir de imágenes de alta resolución y modelos digitales del terreno. Boletín de la SELPER 10: 8-15

Castro R, Chuvieco E (1998) Modeling forest fire danger from geographic information systems. Geocarto International 13: 15-23

Chandler C, Cheney P, Thomas P, Traband L, Williams D (1983) Fire in Forestry. Vol I: Forest fire behavior and effects. John Wiley & Sons, New York

Chavez PS (1996) Image-based atmospheric corrections. Revisited and improved. Photogrammetric Engineering and Remote Sensing 62: 1025-1036

Chavez PS, JR, Sides SC, Anderson JA (1991) Comparison of three different methods to merge multiresolution and multispectral data: Landsat TM and SPOT Panchromatic. Photogrammetric Engineering and Remote Sensing 57: 295-303

Chladil MA, Nunez M (1995) Assessing grassland moisture and biomass in Tasmania. The application of remote sensing and empirical models for a cloudy environment. International Journal of Wildland Fire 5: 165-171

Chou YH (1992) Management of wildfires with a geographical information system. International Journal of Geographical Information Systems 6: 123-140

Chou YH, Minnich RA, Salazar LA, Power JD, Dezzani RJ (1990) Spatial autocorrelation of wildfire distribution in the Idyllwild Quadrangle, San Jacinto mountain, California. Photogrammetric Engineering and Remote Sensing 56: 1507-1513

Christensen NL (1994) The effects of fire on physical and chemical properties of soils in Mediterranean-climate shrublands. In: Moreno JM, Oechel WC (eds) The role of fire in Mediterranean-type Ecosystems. Springer-Verlag, New York, pp 79-95

Churchill PN, Sieber AJ (1991) The current status of ERS-1 and the role of radar remote sensing for the management of natural resources in developing countries. In: Belward AS, Valenzuela CR (eds) Remote sensing and Geographical Information Systems for resource management in developing countries. ECSC, EEC, EAEC, The Netherlands, pp 111-143

Chuvieco, E, (1989) Multitemporal Analysis of Thematic Mapper Images. Applications to Forest Fire Mapping and Inventory in a Mediterranean Environment. Proceedings of a

Workshop on 'Earthnet Pilot Project on Landsat-Thematic Mapper Applications', Frascati, pp 279-285

Chuvieco E (1995) Mapping landscape change in fire-altered areas from remote sensing considering temporal and spatial resolution. In: Ben D, Hubert G (eds) Remote sensing in landscape ecological mapping. Joint Research Center, Institute for remote sensing applications, EUR 16265 EN, pp 101-111

Chuvieco E, Congalton RG (1988) Mapping and inventory of forest fires from digital processing of TM data. Geocarto International 4: 41-53.

Chuvieco E, Congalton RG (1989) Application of remote sensing and Geographic Information Systems to Forest fire hazard mapping. Remote Sensing of Environment 29: 147-159

Chuvieco E, Martín MP (1994) A simple method for fire growth mapping using AVHRR Channel 3 data. International Journal of Remote Sensing 15: 3141-3146

Chuvieco E, Martín, MP (1998) Radiometric analysis to assess the potential performance of the FUEGO System. Final Report for INSA - FUEGO programme. Universidad de Alcalá, Alcalá de Henares

Chuvieco E, Salas FJ (1996) Mapping the spatial distribution of forest fire danger using GIS. International Journal of Geographical Information Systems 10: 333-345

Chuvieco E, Salas FJ, Vega C (1997) Remote sensing and GIS for long-term fire risk mapping, In: Chuvieco E (ed) A review of remote sensing methods for the study of large wildland fires. Universidad de Alcalá, Alcalá de Henares, pp 91-107

Chuvieco E, Salas J, Barredo JI, Carvacho L, Karteris M, Koutsias N (1998) Global patterns of large fire occurrence in the European Mediterran Basin. A G.I.S. analysis. In: Viegas DX (ed) III International Conference on Forest Fire Research - 14th Conference on Fire and Forest Meteorology. ADAI, Luso-Coimbra, pp 2447-2462

Cibula WG, Zetka EF, Rickman DL (1992) Response of Thematic Mapper bands to plant water stress. International Journal Remote Sensing 13: 1869-80

Coffin DP, Lauenroth WK (1990) A gap dynamics simulation-model of succession in a semiarid grassland. Ecol Modelling 49: 229-266

Cohen WB (1991) Temporal versus spatial variation in leaf reflectance under changing water stress conditions. International Journal of Remote Sensing 12: 1865-1876

Coll C, Caselles V (1997) A global split-window algorithm for land surface temperature from AVHRR data: validation and algorithm comparison. Journal of Geophysical Research 102B14:16697-16713

Coll C, Caselles V, Sobrino JA, Valor E (1994) On the atmospheric dependence of the split winclow equation for land surface temperature. International Journal of Remote Sensing 15: 105-122

Conese, C, Maracchi G, Miglietta F, Maselli F, Sacco VM (1998) Forest classification by principal component analysis of TM data. International Journal of Remote Sensing 9: 1597-1612

CONRAC Corp, Conrac Devision (1980) Raster Graphics Handbook, Conrac Corp, Corina

Cope MJ, Chaloner WG (1985) Wildfire: an interaction of biological and physical processes. In: Tiffney BH (ed) Geological factors and the evolution of plants. Yale University Press, New Haven, pp 257-277

Cosentino MJ, Estes JE (1981) Use of Landsat data to develop a fuels database for a wildland fire simulation model. Pecora VII Symposium, Sioux Falls, pp 590-599

Cracknell AP (1997) The Advanced Very High Resolution Radiometer (AVHRR). Taylor & Francis, London

Dagorne A, Dauphiné A, Escleyne G, Gueron L, Baudoin L, Lenco M (1990) L'utilisation de la télédétection aérospatiale en mode multi-satellites, multi-capteurs et multi-dates pour l'étude de la reprise de la végétation après incendie. Photo-Interprétation 5: 45-51

Dagorne A, Duché Y, Castex JM, Ottavi JY (1994) Protection des fôrets contre l'incendie & système d'information géographique. Application à la commune d'Auribeau-sur-Siagne (Alpes-Maritimes). Fôret Méditerranéenne 15: 409-420

Darlington RB (1980) Regression and linear models. McGraw-Hill, New York

DeBano LF, Eberlein GE, Dunn PH (1979) Effects of burning on Chaparral soils: I. Soil nitrogen. Soil Sci Soc Am J 43: 504-509

Deeming JE, Lancaster JW, Fosberg MA, Furman RW, Schroeder MJ (1974) The National Fire-Danger Rating System. RM-84. USDA, Rocky Mountain Forest and Range Experiment Station, Fort Collins

Deeming JE, Burgan RE, Cohen JD (1977) The National Fire-Danger Rating System - 1978. GTR INT-39. USDA Forest Service, Ogdan

Desbois N, Vidal A (1996) Real time monitoring of vegetation flammability using NOAA-AVHRR thermal infrared data. EARSEL Journal Advances in Remote Sensing 4: 25-32

Desbois N, Deshayes M, Beudoin A (1997a) Protocol for fuel moisture content measurements. In: Chuvieco E (ed) A review of remote sensing methods for the study of large wildland fires. Universidad de Alcalá, Alcalá de Henares, pp 61-72

Desbois N, Pereira JMC, Beudoin A, Chuvieco E, Vidal A (1997b) Short term fire risk mapping using remote sensing. In: Chuvieco E (ed) A review of remote sensing methods for the study of large wildland fires. Universidad de Alcalá, Alcalá de Henares, pp 29-60

Dixon R, Shipley R, Briggs A (1984) Landsat - a tool for mapping fuel types in the Boreal Forest of Manitoba. A pilot study. Manitoba Remote Sensing Center. Canada Centre for Remote Sensing, Winnipeg

Domínguez L, Lee B, Chuvieco E, Cihlar J (1994) Fire danger estimation using AVHRR images in the prairie provinces of Canada. II International Conference on Forest Fire Research, Coimbra, pp 679-690

Dowty PR (1993) A theoretical study of fire detection using AVHRR data. M.Sc. thesis, Department of Environmental Sciences, University of Virginia

Dozier J (1981) A method for satellite identification of surface temperature fields of subpixel resolution. Remote Sensing of Environment 11: 221-229.

Dwired RS, Sankar RT (1992) Principal component analysis of Landsat MSS data for delineation of terrain features. International Journal of Remote Sensing 13: 2309-2318

Eastman JR (1993) IDRISI. Clark University, Worcester

Edwards K, Davis PA (1994) The use of intensity-hue-saturation transformation for producing color shaded-relief images. Photogrammetric Engineering and Remote Sensing 60: 1369-1374

EEA (1996) Natural Resources. European Environmental Agency, Copenhagen

Eidenshink JC, Haas RH, Zokaites DM, Ohlen DO, Gallo KP (1989) Integration of remote sensing and GIS technology to monitor fire danger in the Northern Great Plains, Proc. Challenge for the 1990's GIS, Ottawa , pp 944-956

Elvidge CD, Baugh KB, Hobson VR, Kihn E, Kroehl HW, Davis ER, Cocero D (1997) Satellite inventory of human settlements using nocturnal radiation emissions: a contribution for the global toolchest. Global Change Biology 3: 387-395

ESRI (1997) ARC/INFO. ESRI, Redlands

Eva H, Flasse S (1996) Contextual and multi-threshold algorithms for regional active fire detection with AVHRR data. Remote Sensing Reviews 14: 333-351

Eva HD, Belward AS, Grégoire J-M, Moula M , Brustet JM, Janodet E , Viovy N (1995) The application of Along Track Scanning Radiometer to burnt area mapping in different savannah ecosystems in central Africa. Proceedings of the 1995 Meteorological Satellite Data User's Conference, Winchester, pp 201-208

Ferran A (1996) La fertilitat dels sòls forestals en la regeneració després del foc de diferents ecosistemes mediterranis. PhD thesis, Universitat de Barcelona

Ferran A, Vallejo VR (1992) Litter dynamics in post-fire successional forests of *Quercus ilex*. Vegetatio 99: 239-246

Ferran A, Delitti W, Vallejo VR (1998) Effects of different fire recurrences in *Quercus coccifera* communities of the Valencia region. In: Viegas DX (ed) III International Conference on Forest Fire Research - 14th Conference on Fire and Forest Meteorology. ADAI, Luso-Coimbra, pp 1555-1569

Flannigan MD, Vonder Haar TH (1986) Forest fire monitoring using NOAA satellite AVHRR. Canadian Journal of Forest Research 16: 975-982

Flannigan MD, van Wagner CE (1991) Climate change and wildfire in Canada. Can J For Res 21: 66-72

Flasse SP, Ceccato P (1996) A contextual algorithm for AVHRR fire detection. International Journal of Remote Sensing 17: 419-424

Flasse SP, Ceccato P, Downey ID, Raimadoya MA, Navarro P (1997) Remote sensing and GIS tools to support vegetation fire management in developing countries. Proceedings of IGARSS'97, Singapore, pp 1569-1572

Flasse SP, Downey ID, Jacques de Dixmude A, Navarro P, Alvarez R, Zuniga Z, Humphrey I, Uriarte F and Ramos A (1998) Cost-effective operational remote sensing in support of forest fire monitoring: recent experiences from Nicaragua. Proceedings of the 27th International Symposium on Remote Sensing and Environment. Information for Sustainability, Tromsø, pp 773-776

Forman RTT, Gordon M (1986) Landscape Ecology. John Wiley & Sons, New York

Fosberg MA, Sestak ML (1986) KRISSY: user's guide to modelling three dimensional wind flow in complex terrain. USDA, Forest Service, Pacific Southwest Forest and Range Experiment Station, Berkeley

Fosberg MA, Rothermel RC, Andrews PL (1981) Moisture content calculations for 1000-Hour timelag fuels. Forest Science 27: 19-26

Fox BJ, Fox MD (1986) Resilience of animal and plant communities to human disturbance. In: Dell B, Hopkins AJM, Lamont BB (eds) Resilience in Mediterranean-type Ecosystems. Dr W Junk Publishers, Dordrecht, pp 39-64

França JRA, Brustet JM, Fontan J (1995) Multispectral remote sensing of biomass burning in West Africa. Journal of Atmospheric Chemistry 22: 81-110

Frederiksen P, Langaas S, Mbaye M (1990) NOAA AVHRR and GIS-based monitoring of fire activity in Senegal - a provisional methodology and potential applications. In: Goldammer, JG (ed) Fires in tropical biota, Springer-Verlag, Berlin, pp 400-417

French NHF, Kasischke ES, Bourgeau-Chavez LL, Harrell PA, Christensen NL (1994) Relating soil water measurements at fire disturbed sites in Alaska to ERS-1 SAR image signature. Proceedings IGARSS'94, Pasadena, pp.246-248

Fujioka FM (1983) Weighted interpolation as an interpretive tool. Implications for meteorological network design. 7th Conference on fire and forest meteorology. American Meteorological Society, Fort Collins, pp 1-6

Fuller SP, Rouse WR (1979) Spectral reflectance changes accompanying a post-fire recovery sequence in a subartic spruce-lichen woodland. Remote Sensing of Environment 8: 11-23

Fung T, LeDrew E (1987) Application of principal components analysis to change detection. Photogrammetric Engineering and Remote Sensing 53: 1649-1658

García-Ruiz JM, Lasanta, T, Ruiz-Flano P, Ortigosa L, White S, González C, Martí C (1996) Land-use changes and sustainable development in mountain areas: A case study in the Spanish Pyrenees. Landscape Ecol 11: 267-277

Giglio L, Kendall JD, Justice CO (1998) Evaluation of global fire detection algorithms using simulated AVHRR infrared data. Submitted to International Journal of Remote Sensing

Gill AM (1981) Adaptative responses of Australian vascular plant species to fire. In: Gill AM, Groves RH, Noble RI (eds) Fire and the Australian Biota. Australian Academy of Sciences, Canberra, pp 243–272

Gill AM (1981) Adaptative responses of Australian vascular plant species to fire. In: Gill AM, Groves RH, Noble RI (eds) Fire and the Australian Biota. Australian Academy of Sciences, Canberra, pp 243–272

Gillespie AR, Kahle AB, Walker RE (1986) Color enhancement of highly correlated images. I: Decorrelation and HIS contrast stretches. Remote Sensing of Environment 20: 209-235

Gillon D, Rapp M (1989) Nutrient losses during a winter low-intensity prescribed fire in a Mediterranean forest. Plant Soil 120: 69-77

Gluck MJ, Rempel RS (1995) The effect of scale on structural measurements of post-disturbance vegetation in Northwestern Ontario. In: Chuvieco E (ed) Remote Sensing and GIS applications to forest fire management. Universidad de Alcalá, Alcalá de Henares, pp 45-48

González F, Cuevas JM, Casanova JL, Calle A, Illera P (1997) A forest fire risk assessment using NOAA-AVHRR images in the Valencia area, Eastern Spain. International Journal of Remote Sensing 18: 2201-2207

Goodchild MF, Quattrochi PA (1997) Introduction: Scale, multiscaling, Remote sensing and GIS. In: Quattrochi PA, Goodchild MF (eds) Scale in remote sensing and GIS. CRC Lewis publishers, Boca Raton, pp 1-11

Gouma V, Chronopoulou-Sereli A (1998) Wildland fire danger zoning - A methodology. International Journal of Wildland Fire 8: 37-43

Grace J (1983) Plant Water Relationships. Chapman and Hall, London

Green WB (1983) Digital image processing-a systems approach. Van Nostrand Reinhold Co, New York

Grégoire J-M, Belward AS and Kennedy PJ (1993) Dynamique du Saturation du signal dans la bande 3 du senseur AVHRR: Handicap majeur ou source d'information pour la surveillance de l'environnement en milieu soudano-guinéen d'Afrique de l'Ouest?. International Journal of Remote Sensing 14: 2079-2095

Grigoryev AA, Kipatov VR (1976) Space remotes sensing of smokes. Proceedings of the 10th International Symposium on Remote Sensing and Environment. Environmental Research Institute of Michigan, Ann Arbor, pp 305-318

Grove AT (1996) The historical context: before 1850. In: Brandt CJ, Thornes J (eds) Mediterranean desertification and land use. J Wiley & Sons, Chichester, pp 13-28

Gum PW (1985) Computerization of fire dispatch utilising satellite Imagery "Okanogan Project". Pecora 10 Symposium: Remote Sensing in Forest and Range Resource Management, Fort Collins, pp 315-325

Haines DA (1988) A lower atmosphere severity index for wildlife fires. National Weather Digest 13: 23-27

Haines DA, Johnson VJ, Main WA (1976) An assessment of three measures of long-term moisture deficiency before critical fire periods. INT-131. USDA - North Central Forest and Range Experiment Station, St Paul

Hale M, Orcutt DM (1987) The Physiology of Plants Under Stress. John Wiley & Sons, New York

Hall DK, Ormsby JP, Johnson L, Brown J (1980) Landsat digital analysis of the initial recovery of burned tundra at Kokolik River, Alaska. Remote Sensing of Environment 10: 263-272

Harris A (1996) Towards automated fire monitoring from space: semi-automated mapping of the January 1994 New South Wales wildfires using AVHRR data. International Journal of Wildland Fire 6: 107-116

Hartford R, Burgan R (1994) Vegetation condition and fire occurrence: a remote sensing connection. Interior West Fire Council Meeting and Symposium, Coeur d' Alene, pp 1-14

Hartford RA, Rothermel RC (1991) Fuel moisture as measured and predicted during the 1988 Fires in Yellowstone Park. USDA, Forest Service, Missoula

Haydn R, Dalke GW, Henkel J (1982) Application of the HIS color transform to the processing of multisensor data and image enhancement. Proceedings of the International Symposium on Remote Sensing of Arid and Semi-Arid Lands, Cairo, pp 599-616

Helm G, Neal B, Taylor L (1973) A fire hazard severity classification system for California's wildlands. Resources Agency, State of California

Hewitson BC, Crane RG (1994) Neural nets: applications in Geography. Kluwer Academic Publishers, Dordrecht

Hirsch SN, Kruckeberg RF, Madden FH (1971) The bi-spectral forest detection system. 7th International Symposium on Remote Sensing of Environment, Ann Arbor, pp 2253-2259

Hlavka CA, Ambrosia VG, Brass JA, Rezendez AR, Guild LS (1996) Mapping fire scars in the Brazilian cerrado using AVHRR imagery. In: Levine JS (ed) Biomass Burning and Global Change, MIT Press, Cambridge, pp 555-560

Holder G, Wyngaarden V (1990) Flexible analysis through the integration of a fire growth model using an analytical GIS. GIS '90 Symposium, Vancouver, pp 153-157

Hougham AM (1987) Use of NOAA AVHRR digital satellite data for precipitation and forest fire assessment. Technical Report 198. Saskatchewan Research Council , Saskatoon

Houghton JT, Meiro Filho LG, Callander BA, Kattenburg A, Maskell K (eds) (1996) Climate Change 1995. The Second Assessment Report of the IPCC. Cambridge University Press, Cambridge

Hubbard KG (1994) Spatial variability of daily weather variables in the high plains of the USA. Agricultural and Forest Meteorology 68: 29-41

Huete AR (1988) A soil-adjusted vegetation index (SAVI). Remote Sensing of Environment 25: 295-309

Hunt ER, Rock BN (1989) Detection of changes in leaf water content using near and middle-infrared reflectances. Remote Sensing of Environment 30: 43-54

Hunt ER, Rock BN, Nobel PS (1987) Measurement of leaf relative water content by infrared reflectance. Remote Sensing of Environment 22: 429-435

ICONA (1981) Curso superior de defensa contra los incendios forestales, ICONA, Madrid

ICONA (1993) Manual de operaciones contra incendios forestales. ICONA, Madrid

Illera P, Fernández A, Casanova JL (1995) Automatic algorithm for the detection and analysis of fires by means of NOAA AVHRR images. EARSel Advances in Remote Sensing 4: 1-6

Illera P, Fernández A, Delgado JA (1996) Temporal evolution of the NDVI as an indicator of forest fire danger. International Journal of Remote Sensing 17: 1093-1105

Infocarto (1998) Fire Detection, Megafires Final Report (ENV-CT96-0256), University of Alcalá for the European Commission, Alcalá de Henares

Isaacson DL, Smith HG, Alexander CJ (1982) Erosion hazard reduction in a wildfire damaged area. In: Johannsen CJ, Sanders JL (eds) Remote sensing for resource management. Soil Conservation Society of America, Ankeny, pp 179-190

Jackson RD (1986) Remote sensing of biotic and abiotic plant stress. Ann Rev Phytopathology 24: 265-287

Jackson RD, Ezra CE (1985) Spectral response of cotton to suddlenly induced water stress. International Journal Remote Sensing 6: 177-185

Jackson RD, Idso SB, Reginato RJ, Pinter PJ (1981) Canopy temperature as a crop water stress indicator. Water Resources Research 17: 1133-1138

Jakubauskas ME, Lulla KP, Mausel PW (1990) Assessment of vegetation change in a fire-altered forest landscape. Photogrammetric Engineering and Remote Sensing 56: 371-377

Jensen JR (1986) Introductory digital image processing- A remote sensing perspective. Prentice-Hall, Englewood Cliffs

Johnson CE, Thomas LR (1951) The polaroid camera in fire control. Forest Service fire control notes 12: 24-25

Johnson RA, Wichern DW (1998) Applied multivariate statistical analysis. Prentice-Hall, Upper Saddle River

Jones TP, Chaloner WG (1991) Fossil charcoal, its recognition and palaeoatmospheric significance. Palaeogeography, Palaeoclimatology, Palaeoecology 93: 39-50

Justice CO, Dowty P (1994) IGBP-DIS Satellite fire detection algorithm workshop technical report, IGBP-DIS Working Paper #9. NASA/GSFC, Greenbelt

Justice CO, Malingreau J-P, Setzer AW (1993) Satellite remote sensing of fires: potential and limitations. In: Crutzen PJ, Goldammer JG (eds) Fire in the Environment: The ecological, atmospheric, and climatic importance of vegetation fires. John Wiley and Sons, New York, pp 77-88

Kailidis DS (1992) Forest fires in Greece. Proceedings of the International Seminar on Forest Prevention, Land Use and People. Greek Ministry of Agriculture, Athens, pp 27-40

Karteris M (1995) Burned land mapping and post-fire effects. In: Chuvieco E (ed) Remote Sensing and GIS applications to forest fire management. Universidad de Alcalá, Alcalá de Henares, pp 35-44

Kasischke ES, Bourgeau-Chavez LL, French NHF, Harrel P, Christensen Jr NL (1992) Initial observations on using SAR to monitor wildfire scars in boreal forests. International Journal of Remote Sensing 13: 3495-3501

Kasischke ES, French NHF, Harrel P, Christensen Jr NL, Ustin SL, Barry D (1993) Monitoring wildfires in boreal forests using large area AVHRR NDVI composite image data. Remote Sensing of Environment 45: 61-71

Kasischke ES, Bourgeau-Chavez LL, French NHF (1994) Observations of variations in ERS-1 SAR image intensity associated with forest fires in Alaska. IEEE Transactions on Geoscience and Remote Sensing 32: 206-210

Kaufman Y, Tucker CJ, Fung I (1989) Remote sensing of biomass burning in the tropics. Advances in Space Research 9: 265-268

Kaufman YJ, Tucker CJ, Fung I (1990a) Remote sensing of biomass burning in the tropics. Journal of Geophysical Research 95: 9927-9939

Kaufman YJ, Setzer A, Justice C, Tucker CJ, Pereira MC, Fung I (1990b) Remote sensing of biomass burning in the tropics. In: Goldammer JG (ed) Fires in tropical biota. Springer-Verlag, Berlin, pp 371-399

Kaufman YJ, Setzer A, Ward D, Tanre D, Holben BN, Menzel P, Rasmussen R (1992) Biomass burning airborne and space-borne experiment in the Amazons (BASE-A). Journal of Geophysical Research 97: 14581-14599

Kaufman YJ, Tanré D (1992) Atmospherically Resistant Vegetation Index (ARVI) for EOS-MODIS. IEE Transactions on Geoscience and Remote Sensing 30: 261-270

Keetch JJ, Byram GM (1968) A drought index for forest fire control. USDA Forest Service, SE-38, pp 32

Kennedy PJ (1992) Biomass burning studies: the use of remote sensing. Ecological Bulletins 42:133-148

Kennedy PJ, Belward AS, Grégoire JM (1994) An improved approach to fire monitoring in West Africa using AVHRR data. International Journal of Remote Sensing 15: 2235-2255

Khanna PK, Raison RJ (1986) Effects of fire intensity on solution chemistry of surface soil under Eucalyptus pauciflora forest. Aust J Soil Res 24: 423-434

Kneppeck ID, Ahern FJ (1989) Stratification of a regenerating burned forest in Alberta using Thematic Mapper and C-SAR images. Proceeding, IGARSS'89 Symposium, pp 1391-1396

Koffi B, Grégoire J-M, Mahé G, Lacaux J-P (1995) Remote sensing of bush fire dynamics in Central Africa from 1984 to 1988: analysis in relation to regional vegetation and pluviometric patterns. Journal of Atmospheric Research 39: 179-200

Koffi B, Grégoire J-M, Mahé G, Lacaux J-P (1995) Remote sensing of bush fire dynamics in Central Africa from 1984 to 1988: analysis in relation to regional vegetation and pluviometric patterns. Journal of Atmospheric Research 39: 179-200

Köppen W (1936) Das geographische System der Klimate Handbuch der Klimatologie. Teil C, Berlin

Kosmas CS (1996) Impacts of desertification and agriculture. International Conference on Mediterranean Desertification. Research Results and Policy Implications. Crete-Hellas, pp. 16-17

Kourtz PH (1977) An application of Landsat digital technology to forest fire fuel type mapping. 11th International Symposium on Remote Sensing of Environment, Ann Arbor, pp 1111-1115

Koutsias N, Karteris M (1996) Logistic regression modelling of Thematic Mapper data for burnt area mapping. Unpublished manuscript, Aristotelian University of Thessaloniki, Dept. of Forestry and Natural Environment, Lab. of Forest Management and Remote Sensing

Koutsias N, Karteris M (1998a) Logistic regression modelling of multitemporal Thematic Mapper data for burned area mapping. International Journal of Remote Sensing 19: 3499-3514

Koutsias N, Karteris M (1998b) Burned area mapping using logistic regression modeling of a single post-fire Landsat-5 Thematic Mapper image. Submitted to International Journal of Remote Sensing

Koutsias N, Karteris M, Chuvieco E (1998) The use of intensity-hue-saturation transformation of Landsat-5 Thematic Mapper data for burned land mapping. Photogrametric Engineering and Remote Sensing (in preparation)

Kozlowski TT, Kramer PJ, Pallardy S (1991) The Physiological Ecology of Woody Plants. Academic Press, San Diego

Kutiel P, Naveh Z (1987) The effect of fire on nutrients in a pine forest soil. Plant Soil 104: 269-274

Laguna E, Reyna S (1990) Diferencias entre los óptimos natural y forestal de las vegetaciones valencianas y alternativas futuras de gestión. Ecología, Fuera de Serie 1: 321-330

Lam N, Quattrochi DA (1992) On the issues of scale, resolution and fractal analysis in the mapping sciences. Professional Geographer 44: 88-98

Landry R, Ahern FJ, O'Neil R (1995) Forest burn visibility on C-HH radar images. Canadian Journal of Remote Sensing 21: 204-206

Langaas S (1989) A study of spectral characteristics of burnt areas in Senegal, with recommendations for an AVHRR based bush fire monitoring methodology, Dept. of Surveying, Agric. Univ. of Norway, Aas.

Langaas S (1992) Temporal and spatial distribution of savannah fires in Senegal and the Gambia, West Africa, 1989-90, derived from multi-temporal AVHRR night images. International Journal of Wildland Fire 2: 21-36

Langaas S (1993) A parameterised bispectral model for savannah fire detection using AVHRR night images. International Journal of Remote Sensing 14: 2245-2262

Langaas S (1995) Night-time observations of West-African bushfires from space. PhD Thesis, Department of Geography, University of Oslo

Langaas S, Kane R (1991) Temporal spectral signatures of fire scars in savanna woodland. Proceedings IGARSS'91 Remote Sensing: Global Monitoring for Earth Management. IEEE Publications, Espoo, pp 1157-1160

Langaas S, Muirhead (1989) Monitoring of bush fires in West-Africa by weather satellites. Proceedings of the 22 International Symposium on Remote Sensing of Environment, Abidjan, pp 253-268

Langhart R, Bachman A, B A (1998) Temporal and spatial patterns of wildfire occurrence (Canton de Grison, Switzerland). In: Viegas DX (ed) III International

Lasaponara R, Cuomo V and Tramutoly V (1998) Satellite forest fire detection in the Italian ecosystems using AVHRR data. In: Viegas DX (ed) III International Conference on Forest Fire Research - 14th Conference on Fire and Forest Meteorology. ADAI, Luso-Coimbra, pp 2013-2028

Le Houérou HN (1993) Land degradation in Mediterranean Europe: can agroforestry be a part of the solution? A prospective review. Agrofor Sys 21: 43-61

Lee TF, Tag PM (1990) Improved detection of hotspots using the AVHRR 3.7-µm channel. Bulletin of the American Meteorological Society 71:1722-1730

Lillesand TM, Kiefer RW (1994) Remote sensing and image interpretation. John Wiley and Sons, New York

Little SN, Ohmann JL (1988) Estimating nitrogen lost from forest floor during prescribed fires in Douglas-Fir Western Hemlock clearcuts. Forest Sci 34: 152-164

Lobo A, Navarro R, Pineda N, Fernández P, Salas FJ, Fernánez JL, Fernández-Palacios A (1998). Assessment of Fire Impact in Mediterranean Forests Based on Landsat Thematic Mapper Imagery. Report to the Institute of Space Applications. Joint Research Centre, European Commission. Unpublished

Loftsgaarden D, Andrews PL (1992) Constructing and testing logistic regression models for binary data: applications to the National Fire Danger Rating System. GTR INT-286. USDA, Forest Service, Ogden

Lombrana MJ (1995) Monitoring of burnt forest areas with remote sensing data. A study in North-East Spain using Landsat TM and Spot XS data. Technical note I.95.80. Institute for Remote sensing applications, Joint Research Centre, Ispra

López G, Caselles V (1991) Mapping burns and natural reforestation using Thematic Mapper data. Geocarto International 6: 31-37

López S, González F, Llop R, Cuevas JM (1991) An evaluation of the utility of NOAA AVHRR images for monitoring forest fire risk in Spain. International Journal of Remote Sensing 12: 1841-1851

Lu J, Bomba P, Kind T (1990) A microcomputer-based geographic information system for forest fire management. ACSM-ASPRS Annual Convention, Denver, pp 180-192

Maheras P (1988) Changes in precipitation conditions in the western Mediterranean over the last century. J Clim 8: 179-189

Malingreau J-P, Tucker CJ (1988) Large scale deforestation in the south-eastern Amazon basin of Brazil. Ambio 17: 49-55

Malingreau J-P (1990) The contribution of remote sensing to the global monitoring of fires in tropical and subtropical ecosystems. In: Goldammer JG (ed) Fire in Tropical Biota. Springer Verlag, Berlin, pp 337-370

Malingreau J-P, Leysen M, Degrandi F (1995) Detecting and measuring burn scars in tropical vegetation using ERS-1 SAR data. IGBP-DIS Workshop on Global Fire Monitoring. Joint Research Centre, Ispra

Mandallaz D, Ye R (1997) Prediction of forest fires with Poisson models. Canadian Journal of Forest Research 27: 1685-1694

Marchetti M (1990) Forest Fire Hazard Map: Pilot Project For Italy. 1st Int Conf Forest Fire Research, Coimbra, pp B.25: 1-12

Margaris NS, Koutsidou E, Giourga Ch (1996) Changes in traditional Mediterranean land-use systems. In: Brandt CJ, Thornes J (eds) Mediterranean desertification and land use. J Wiley & Sons, Chichester, pp 29-42

Martell DL, Otukol S, Stocks BJ (1987) A logistic model for predicting daily people-caused fire occurrence in Ontario. Canadian Journal of Forest Research 17: 394-401

Martín MP, Viedma D, Chuvieco E (1994) High versus low resolution satellite images to estimate burned areas in large forest fires. 2nd Int Conf Forest Fire Research, Coimbra, pp 653-663

Martínez-Ruiz E (1994) El problema de los incendios forestales en España, análisis de los últimos cincuenta años, previsones de cara al siglo XXI. Rev For Española 11: 40-54

Mather PM (1987) Computer Processing of Remotely-Sensed Images. An Introduction. John Willey & Sons, New York

Matson M, Dozier J (1981) Identification of sub resolution high temperature sources using a thermal IR sensor. Photogrammetric Engineering and Remote Sensing 47: 1311-1318

Matson M, Holben B (1987) Satellite detection of tropical burning in Brazil. International Journal of Remote Sensing 8: 961-970

McArthur AG (1966) Weather and grassland fire behaviour. Report 100, Commonwealth of Australia Forest and Timber Bureau, Canberra

McArthur AG (1967) Fire behaviour in eucalypt forests. Report 107, Commonwealth of Australia Forest and Timber Bureau, Canberra

McCutchan MH, Fox DG (1986) Effect of elevation and aspect on wind, temperature and humidity. Journal of Climate and Applied Meteorology 25: 1996-2013

Menaut JC, Abbadie L, Vitousek PM (1993) Nutrient and organic matter dynamics in Tropical ecosystems. In: Crutzen PJ, Goldammer JG (eds) Fire in the Environment. The Ecological, Atmospheric and Climatic Importance of Vegetation Fires. John Wiley & Sons, Chichester, pp 215-231

Mendenhall W, Sincich T (1996) A second course in statistics: regression analysis. Prentice-Hall, New Jersey

Miller WA, Johnston DC (1985) Comparison of fire fuel maps produced using MSS and AVHRR data. Pecora 10 Symposium. Remote sensing in forest and range resource management. US Geological survey, Fort Collins, pp 305-314

Miller WA, Howard SM, Moore DG (1986) Use of AVHRR data in an information system for fire management in the Western United States. 20th International Symposium on Remote Sensing of Environment, Nairobi, pp 67-79

Milne AK (1986) The use of remote sensing in mapping and monitoring vegetational change associated with bushfire events in Eastern Australia. Geocarto International 1: 25-35

Milne AK, Hall AM (1992) Towards the operational detection and mapping of bushfires. 6th Australasian Remote Sensing Conference, Wellington, pp 156-165

Minick GR, Shain WA (1981) Comparison of satellite imagery and conventional aerial photography in evaluating a large forest fire. Proc 7th International symposium on machine processing of remotely sensed data, West Lafayette, pp 544-546

Moran MS, Clarke TR, Inoue Y, Vidal A (1994) Estimating crop water deficit using the relation between surface-air temperature and spectral vegetation index. Remote Sensing of Environment 49: 246-263

Moreno JM (1998) Large Fires. Backhuys Publishers, Leiden

Moreno JM, Oechel WC (1994) The Role of Fire in Mediterranean-Type Ecosystems. Springer-Verlag, New York

Moreno JM, Vázquez A, Vélez R (1998) Recent history of forest fires in Spain. In: Moreno JM (ed) Large Fires. Backhuys Publishers, Leiden, pp 159-185

Moro C (1996) Variation spatio-temporelle de l'indice de siccité de la bruyère arborescente et de l'arbousier dans le massif des Maures, campagnes des étés 1994 1995 et 1996. INRA-PIF, Avignon

Moro C, Valette JC, (1996) Teneur en eau de combustibles forestiers méditerranéens. INRA-PIF, Avignon

Muirhead K, Cracknell A (1985) Straw burning over Great Britain detected by AVHRR. International Journal of Remote Sensing 6: 827-833

Narumalani S, Jensen JR, Althausen JD, Burkhalter S, Mackey HE (1997) Aquatic macrophyte modeling using GIS and logistic multiple regression. Photogrammetric Engineering and Remote Sensing 63: 41-49

Navarro, C., (1991). Uso de las imágenes Landsat-TM en un sistema de evaluación de daños causados por incendios forestales. Actas de la IV Reunión Científica de la Asociación Española de Teledetección, Sevilla, pp. 40-45.

Navarro, C., (1991). Uso de las imágenes Landsat-TM en un sistema de evaluación de daños causados por incendios forestales. Actas de la IV Reunión Científica de la Asociación Española de Teledetección, Sevilla, pp. 40-45.

Navarro, R., Salas, F.J., Navarro, C., Fernández, P., González, M.P. (1997a) Evaluación de daños producidos por incendio y regeneración posterior de la vegetación. Aplicación de imágenes Landsat-TM a su caracterización y seguimiento, Internal Document, Consejería de Medio Ambiente, Junta de Andalucía, Sevilla

Navarro, R., Salas, F.J., Navarro, C., Fernández, P., González, M.P., Fernández-Palacios, A. and Rodrígez y Silva, F. (1997b) Desarrollo de modelos de evaluación de la regeneración para cubiertas de vegetación después de un incendio. Análisis y evaluación del estado fitosanitario de la vegetación natural mediante sensores remotos, Internal Report, Consejeria de Medio Ambiente, Junta de Andalucía, Sevilla

Naveh Z (1975) The evolutionary significance of fire in the Mediterranean region. Vegetatio 29: 199-208

Ne'eman G, Meir I, Ne'eman R (1993) The effect of ash on the germination and early growth of shoots and roots of *Pinus, Cistus* and annuals. Seed Sci Techn 21: 339-349

Nemani RR, Pierce L, Running SW, Edward S (1993) Developing satellite-derived estimates of surface moisture status. Journal of Applied Meteorology 32: 548-557

Nesterov VG (1949) Combustibility of the forest and methods for its determination. State Industry Press, USSR

Ni W, Li X, Woodcock CE, Caetano MR, Strahler A (1998) An analytical hybrid GORT model for bidirectional reflectance over discontinuous plant canopies. Submitted to IEEE Transactions on Geosciences and Remote Sensing

Noble IR, Bary GAV, Gill AM (1980) McArthur's fire-danger meters expressed as equations. Australian Journal of Ecology 5: 201-203

Norusis MJ (1990) SPSS/PC+ Advanced Statistics[TM] 4.0 for the IBM PC/XT/AT and PS/2. SPSS Inc, Chicago, pp 39-61

Norusis MJ (1993) SPSS for Windows. Advanced Statistics Release 6.0. SPSS Inc, Chicago

O'Connell AM (1989) Nutrient accumulation in and release from the litter layer of karri (*Eucaliptus diversicolor*) forests of Southwestern Australia. For Ecol Manage 26: 95-111

Olson CM (1980) An evaluation of the Keetch-Byram drought index as a predictor of foliar moisture content in a chaparral community. 6th National Conference on Fire and Forest Meteorology. Society of American Foresters, Seattle, pp 241-244

Openshaw S (1994) Neuroclassification of spatial data. In: Hewitson B, Crane -R (eds) Neural Nets: Applications in Geography. Kluwer Academic Publishers, Dordrecht , pp 53-70

Palmieri S, Inghilesi R, Siani A, Martellacci C (1992) Un indice meteorologico di rischio per incendi boschivi. Bollettino Geofisico 15: 49-62

Paltridge GW, Barber J (1988) Monitoring grassland dryness and fire potential in Australia with NOAA/AVHRR data. Remote Sensing of Environment 25: 381-394

Paltridge GW, Mitchell RM (1990) Atmospheric and viewing angle correction of vegetation indices and grassland fuel moisture content derived from NOAA/AVHRR. Remote Sensing of Environment 31: 121-135

Pausas JG (1997) Resprouting of cork-oak (*Quercus suber* L) after fire in NE Spain. J Veg Sci 8: 703-706

Pausas JG (1998) Modelling fire-prone vegetation dynamics. In: Trabaud L (ed) Fire Management and Landscape Ecology. Internat Ass Wildland Fire, Washington, pp 327-334

Pausas JG (1999a) Mediterranean vegetation dynamics: modelling problems and functional types, 140: 27-39.

Pausas JG (1999b) The response of plant functional types to changes in the fire regime in Mediterranean ecosystems. A simulation approach, Journal of Vegetation Science (in press)

Pausas JG, Austin MP, Noble IR (1997) A forest simulation model for predicting eucalypt dynamics and habitat quality for arboreal marsupials. Ecol Applications 7: 921-933

Pausas JG, Austin MP (1998) Potential impact of harvesting for the long-term conservation of arboreal marsupials. Landscape Ecol 13: 103-109

Pausas JG, Carbó E, Gil JM (1999) Post-fire regeneration patterns in Eastern Iberian Peninsula, unpublished report.

Pereira JMC (1992) Burned area mapping with conventional and selective principal component analysis. Finisterra 27: 61-76

Pereira JMC (1999) A comparative evaluation of NOAA/AVHRR vegetation indices for burned surface detection and mapping. IEEE Transactions on Geosciences and Remote Sensing, 37:

Pereira JMC, Vasconcelos MJ (1990) Fire propagation modeling in heterogeneous environments and a new spread algorithm for Firemap. 1st International Conference on Forest Fire Research, Coimbra, pp B.14: 1-15

Pereira JMC, Chuvieco E, Beaudoin A, Desbois N (1997) Remote sensing of burned areas. In: Chuvieco E (ed) A review of remote sensing methods for the study of large wildland fires. Universidad de Alcalá, Alcalá de Henares, pp 127-183

Pereira MC, Setzer AW (1993a) Spectral characteristics of deforestation fires in NOAA/AVHRR images. International Journal of Remote Sensing 14: 583-597

Pereira MC, Setzer AW (1993b) Spectral characteristics of fire scars in Landsat-5 TM images of Amazonia. International Journal of Remote Sensing 14: 2061-2078

Pérez-Ramos B (1997) Factores que controlan la variabilidad espacial de la respuesta de la vegetación al fuego en la sierra de Gredos: Usos del territorio e intensidad del fuego. Ph D thesis, Universidad Complutense de Madrid

Perrin M and Millington A (1997) Monitoring biomass burning with ATSR-2. Proceedings of the 3rd European Remote Sensing Symposium, Florence

Phillips C, NicKey B (1979) The Concept of "Spatial Risk" and its Application to Fire Prevention. Fire Management Notes 39: 7-8

Pinty B, Verstraete MM, (1992) GEMI: a non-linear index to monitor global vegetation from satellites. Vegetatio 101: 15-20

Piñol J, Terradas J (1996) Els incendis forestals al litoral mediterrani catalano-valencià en el període 1968-1995. Butlletí del Centre d'Estudis Selvatans 4:69-96

Piñol J, Terradas J, Lloret F (1998) Climatic warming hazard, and wildfire occurrence in coastal eastern Spain. Clim Change 38: 345-357

Pons X, Solé-Sugrañes L (1994) A Simple Radiometric Correction Model to Improve Automatic Mapping of Vegetation from Multispectral Satellite Data. Remote Sensing of Environment 48: 191-204

Ponzoni FJ, Lee DCL, Filho PH (1986) Assessment of area burnt and vegetation recovery at Brasilia National Park, using Landsat TM data. IV Simpósio Brasileiro de Sensoriamento Remoto, ,Vol.1,INPE, São José dos Campos, pp. 615-621

Pozo D, Olmo FJ, Alados-Arboledas L (1997) Fire detection and growth monitoring using a multitemporal technique on AVHRR mid-infrared and thermal channels. Remote Sensing of Environment 60: 111-120

Prins EM, Menzel WP (1994) Trends in South American biomass burning detected with GOES visible infrared spin scan radiometer atmospheric sounder from 1983 to 1991. Journal of Geophysical Research 99: 16719-16735

Prodon R, Fons R, Athias-Binche F (1987) The impact of fire on animal communities in Mediterranean area. In: Trabaud L (ed) The role of fire in ecological systems. SPB Acad Publ, The Hague, pp121-157

Prosper-Laget V, Douguédroit A, Guinot JP (1994) Mapping the risk of forest fire departure using NOAA satellite information. Int Workshop on Satellite Technology and GIS for Mediterranean Forest Mapping and Fire Management, Thessaloniki, pp 151-163

Puigdefábregas J, Mendizabal T (1998) Perspectives on desertification: western Mediterranean. J Arid Envir 39: 209-224

Rabii HA (1979) An investigation of the utility of Landsat-2 MSS data to the fire-danger rating area, and forest fuel analysis within Crater Lake National Park, Ph.D. dissertation. Oregon State University

Raison RJ, Khanna PK, Woods PV (1985) Mechanisms of element transfer to the atmosphere during vegetation fires. Can J For Res 15: 132-140

Rego FC (1992) Land use changes and wildfires. In: Teller A, Mathy P, Jeffers JNR (eds) Response of forest fires to environmental change. Elsevier, London, pp 367-373

Richards JA (1984) Thematic mapping from multitemporal image data using the principal components transformation. Remote Sensing of Environment 16: 35-46

Richards JA (1986) Remote sensing digital analysis. Springer-Verlag, Berlin

Riggan PJ, Franklin SE, Brass JA, Brooks FE (1994) Perspectives on fire management in Mediterranean ecosystems of southern California. In: Moreno JM, Oechel WC (eds) The Role of Fire in Mediterranean-Type Ecosystems. Springer-Verlag, New York, pp 140-162

Ripple WJ (1986) Spectral reflectance relationships to leaf water stress. Photogrammetric Engineering and Remote Sensing 52: 1669-1675

Robinson JM (1991) Fire from space: global fire evaluation using infrared remote sensing. International Journal of Remote Sensing 12: 3-24

Robinson JM (1991) Problems in global fire evaluation: is remote sensing the solution? In: Levine JS (ed) Global Biomass Burning: Atmospheric, Climatic, and Biospheric Implications. The MIT Press, Cambridge, pp 67-73

Rock BN, Vogelmann JE, Williams DL, Vogelmann AF, Hoshizaki T (1986) Remote detection of forest damage. Bioscience 36: 439-445

Rodríguez y Silva F (1993) Modelos de gestión ordenada de la defensa contra los incendios forestales en los espacios naturales protegidos. Congreso Forestal Español, pp 239-244

Rodríguez y Silva F (1994) Las medidas preventivas en la defensa contra los incendios forestales. III Jornadas sobre Incendios Forestales, Córdoba, pp. 120-136

Rodríguez y Silva F (1995) Modelos de comportamiento del fuego aplicados a la ordenación de áreas forestales. Actas del Taller Internacional sobre Prognosis y Gestión en Incendios Forestales, Proyecto Fondef F-I-13, Santiago de Chile, pp. 166-181

Rodríguez y Silva F (1998a) Local evaluation of the forest fires risk through danger indices, application to the forest regions of Andalusia. In: Viegas DX (ed) III International Conference on Forest Fire Research - 14th Conference on Fire and Forest Meteorology. ADAI, Luso-Coimbra, pp 1071-1084

Rodríguez y Silva F (1998b) Modelo matemático para la determinación de los parámetros meteorológicos necesarios para el seguimiento del riesgo y obtención de los pronósticos de comportamiento del fuego en los incendios forestales. Aplicación al Parque Natural de los Alcornocales. Ph D Thesis. Universidad Politécnica de Madrid. E.T.S. Ingenieros de Montes

Rodríguez y Silva F, Navarro R, Navarro C, González MP (1997) Evaluation of Forest Fire Damage with Landsat-TM and Ancillary Information. In: Chuvieco E (ed) A review of remote sensing methods for the study of large wildland fires. Universidad de Alcalá, Alcalá de Henares, pp 185-192

Root RR, Stitt SCF, Nyquist MO, G.S. W, Agee JK (1985) Vegetation and fire fuel models mapping of North Cascades National Park. Pecora 10 Symposium: Remote Sensing and Range Resource Management, Fort Collins, pp 287-294

Root RR, Stitt SCF, Nyquist MO, G.S. W, Agee JK (1986) Vegetation and fire fuel models mapping of North Cascades National Park, ACSM-ASPRS Annual Convention, Washington, D.C. pp 75-85

Rothermel RC (1972) A mathematical model for predicting fire spread in wildland fuels. Research Paper INT-115, USDA, Forest Service, Ogden

Rothermel RC (1978) A concept for appraising fire in nonuniform fuels. USDA, Forest Service, Intermountain Forest & Range Expt Sta, Missoula

Rothermel RC (1983) How to predict the spread and intensity of forest and range fires. USDA, Forest Service, Ogden

Rothermel RC, Wilson RA, Morris GA, Sackett SS (1986) Modelling moisture content of fine dead wildland fuels: input to BEHAVE fire prediction system. Res Pap INT-359. USDA For Ser Intermountain Research Station, Ogden

Rouse JW, Haas RH, Schell JA, Deering DW, Harlan JC (1974) Monitoring the vernal advancement and retroradiation of natural vegetation. Report RSC 1978-4, Remote Sensing Center, Texas A&M Univ, College Station

Roxo MJ, Mourao JM (1995) Land degradation in south interior Alentejo-Mértola. Historical background of human impact by agriculture. Proceedings of the Conference on Erosion and Land Degradation in the Mediterranean, Aveiro, p. 475

Roxo MJ, Cortesao Casimiro P, Soeiro de Brito R (1996) Inner Lower Alentejo field site: Cereal cropping, soil degradation and desertification. In: Brandt CJ, Thornes J (eds) Mediterranean desertification and land use. J Wiley & Sons, Chichester, pp 111-135

Ryan BC (1983) WNDCOM: estimating surface winds in mountainous terrain. USDA, Forest Service, Pacific Southwest Forest and Range Experiment Station, Berkeley

Ryan KC, Reinhardt ED (1988) Predicting postfire mortality of seven western conifers. Canadian Journal of Forest Research 18: 1291-1297

Sala M, Rubio JL (1994) Soil erosion and degradation as a consequence of forest fires. Geforma Ediciones, Logroño

Salas FJ, Chuvieco E (1994) G.I.S. applications to forest fire risk mapping. Wildfire 3: 7-13

Salas FJ, Viegas MT, Kyun IA, Chuvieco E, Viegas DX (1994) A local risk map for the council of Poiares, Central Portugal: comparison of GIS and field work methods. II International Conference on Forest Fire Research, Coimbra, pp 691-702

Salas FJ, Chuvieco E (1995) Aplicación de imágenes Landsat-TM a la cartografía de modelos combustibles. Revista de Teledetección 5: 18-28

Salas FJ, Navarro R, Navarro C, González, MP, Fernández P (1997) Evaluación de Daños por Incendio y Regeneración Posterior de la Vegetación. Aplicación de Imágenes Landsat-TM a su caracterización y Seguimiento. Junta de Andalucía, Consejería de Medio Ambiente. Unpublished.

Salazar LA (1982) Remote Sensing Techniques Aid in Preattack Planning for Fire Management. USDA, Forest Service, Berkeley

Salazar LA, Soto-Estrada RM, Rechel JL (1990) Using GIS Technology to Define Wildfire Risk in Morelos, Mexico. GIS/LIS '90, Anaheim, pp 645-653

Salvador R, Pons X (1995) The role of fire in a Mediterranean area as a long term landscape modifier. In: Chuvieco E (ed) Remote Sensing and GIS applications to forest fire management. Universidad de Alcalá, Alcalá de Henares, pp 69-71

Serrasolsas I (1994) Fertilitat de sòls forestals afectats pel foc: dinàmica del nitrogen i del fòsfor, Ph D Thesis, Universitat de Barcelona

Setzer AW, Pereira MC (1991a) Amazonia biomass burnings in 1987 and an estimate of their tropospheric emissions. Ambio 20: 19-23

Setzer AW, Pereira MC (1991b) Operational detection of fires in Brazil with NOAA-AVHRR. 24th International Symposium on Remote Sensing of Environment, Rio de Janeiro, pp 469-482

Setzer AW, Malingreau J-P (1993) Temporal variation in the detection limit of fires in NOAA/AVHRR images. Proceedings 6th AVHRR Data Users Meeting, Belgirate, pp 575-579

Setzer AW, Verstraete MM (1994) Fire and glint in AVHRR's Channel 3: a possible reason for the no saturation mystery. International Journal of Remote Sensing 15: 711-718

Sharma S (1996) Applied multivariate techniques. John Wiley & Sons, New York

Shasby MB, Burgan RR, Johnson GR (1981) Broad area forest fuels and topography mapping using digital Landsat and terrain data. Machine Processing of Remotely Sensed Data Symposium, West Lafayette, pp 529-538

Shugart HH (1984) A Theory of Forest Dynamics. Springer-Verlag, New York

Siljeström P, Moreno A (1995) Monitoring burnt areas by principal components analysis of multitemporal TM data. International Journal of Remote Sensing 16: 1577-1587

Silva JMN (1996) Comparing the vegetation indices NDVI and VI7 for burnt area mapping with Landsat 5 TM imagery. Unpublished final report. Instituto Superior de Agronomia, Universidade Técnica de Lisboa (in Portuguese).

Simard AJ (1968) The moisture content of forest fuels - A review of the basic concepts. FF-X-14, Forest Fire Research Institute, Ottawa

Simard AJ, Eenigenburg JE, Hobrla SL (1987) Predicting extreme fire potential. 9th Conference on Fire and Forest Meteorology, San Diego, pp 148-157

Simard AJ, Eenigenburg JE, Main WA (1989) A Weather-Based Fire Season Model. 10th Conference on Fire and Forest meteorology, Ottawa, pp 213 -221

Simard AJ, Eenigenburg JE (1991) Forecasting Fire Severity. 11th Conference on Fire and Forest Meteorology, Missoula, pp 132-140

Smith RB and Woodgate PW (1985) Appraisal of fire damage and inventory for timber salvage by remote sensing in mountain ash forests in Victoria. Australian Forestry 48: 252-263

Smith GM, Vaughan RA (1991) A heat source monitoring system and ita application to straw burning in the UK. Spatial Data 2000 Conference, Oxford University, Oxford, pp 293-305

Sol B (1990) Estimation du risque météorologique d'incendies de forêts dans le sud-est de la France. Revue Forestière Française 42: 263-271

Solomon AM (1986) Transient response of forest to CO_2-induced climate change: simulation modelling experiments in eastern North America. Oecologia 68: 567-579

Sousa AMO (1998) Development of a methodology for mapping burnt areas larger than 500ha in the Iberian Peninsula with AVHRR data. Unpublished Masters Thesis. Department of Civil Engineering, Instituto Superior Técnico, Universidade Técnica de Lisboa (in Portuguese).

SPSS (1995) SPSS for Windows, Users guide

Stewart OC (1956) Fire as a first great force employed by man. In: Thomas WI (ed) Man's role in changing the face of the earth. University of Chicago, Chicago, pp 115-133

Stocks BJ, Lawson BD, Alexander ME, Van Wagner CE, McAlpine RS, Lynham TJ, Dubé DE (1989) The Canadian Forest Fire Danger Rating System: an overview. The Forestry Chronicle 65: 450-457

Strasser MJ, Pausas JG, Noble IR (1996) Modelling the response of eucalypts to fire in the Brindabella Range, ACT. Aust J Ecol 21: 341-344

Stuttard M, Boardman S, Ceccato P, Downey I, Flasse SP, Gooding R, Muirhead K (1995) Global Vegetation Fire Product. Final Report for JRC Contract 10444-94-09-FIEP ISP GB, EOS-95/125-FR-001

Tanaka S, Kimura H, Suga Y (1983) Preparation of a 1:25000 Landsat map for assessment of burnt area on Etajima Island. International Journal of Remote Sensing 4: 17-31

Tartaglini N (1992) Post-fire succession and mycorrhizae in two mediterranean communities. In: Trabaud L, Prodon R (eds) Fire in Mediterranean Ecosystems, CEC Ecosystem Res Rep 5, Brussels, pp 395-402

Terradas J (1987) Ecosistemes terrestres. La resposta als incendis i a altres perturbacions. Diputació de Barcelona, Barcelona

Thanos CA, Georghiou K, Kadis CC, Pantazi C (1992) Cistaceae: A plant family with hard seeds. Israel J Bot 41: 251-263

Thomas JR, Namken LN, Oerther GF, Brown RG (1971) Estimating leaf water content by reflectance measurements. Agronomy Journal 63: 845-847

Thornes J (1990) The interaction of erosion and vegetation dynamics in land degradation: spatial outcomes. In: Thornes J (ed) Vegetation and Erosion. Wiley, New York, pp 41-54

Torn MS, Fried JS (1992) Predicting the impacts of global warming on wildland fire. Clim Change 21: 257-274

Torres P, Honrubia M (1997) Changes and effects of a natural fire on ectomycorrhizal inoculum potential of soil in a *Pinus halepensis* forest. For Ecol Manage 96: 189-196

Trabaud L (1991) Fire regimes and phytomass growth dynamics in a *Quercus coccifera* garrigue. J Veg Sci 2: 307-314

Trabaud L, Campant C (1991) Difficulté de recolonisation naturelle du Pin de Salzmann *Pinus nigra* Arn. ssp. *salzmannii* (Dunal) Franco après incendie. Biol Cons 58: 329-343

Trabaud L (1992) Influence du régime des feux sur les modifications à court terme et la stabilité à long terme de la flore d'une garrigue de *Quercus coccifera*. Rev Ecol (Terre Vie) 47: 209-230

Trabaud L (1994) The effect of fire on nutrient losses and cycling in a *Quercus coccifera* garrigue (southern France). Oecologia 99: 379-386

Trabaud L, Galtié JF (1996) Effects of fire frequency on plant communities and landscape pattern in the massif des Aspres (southern France). Landscape Ecol 11: 215-224

Trimble SW (1988) The impact of organisms on overall erosion rates within catchments in temperate regions. In: Viles HA (ed) Biogeomorphology. Blackwell, Oxford, pp 83-142

Tsitsoni T (1997) Conditions determining natural regeneration after wildfires in the *Pinus halepensis* (Miller, 1768) forests of Kassandra Peninsula (North Greece). For Ecol Manage 92: 199-208

Tucker CJ (1979) Red and photographic infrared linear combinations for monitoring vegetation. Remote Sensing of Environment 18: 127-150

Tucker CJ (1980) Remote sensing of leaf water content in the near infrared. Remote Sensing of Environment 10: 23-32

Tucker CJ, Newcomb WW and Dregne HE (1994) AVHRR data sets for the determination of desert spatial extent. International Journal of Remote Sensing 15: 3547-3566

Ustin SL, Smith MO, Adams JB (1993) Remote sensing of ecological processes: a strategy for developing and testing ecological models using spectral mixture analysis. In: Ehrlinger JR, Field CB (eds) Scaling physiological processes: Leaf to Globe. Academic Press, San Diego, pp 339-357

Valette JC (1990) Inflammabilités des espèces forestières méditerranéennes. Revue Forestière Française 42: 76-92

Vallejo VR (1997) La restauración de la cubierta vegetal en la Comunidad Valenciana. Fundación CEAM, Valencia

Vallejo VR, Alloza JA (1998) The restoration of burned lands: The case of eastern Spain. In: Moreno JM (ed) Large forest fires, Backhuys Publ, Leiden, pp 91-108

Vallejo VR, Bautista S, Cortina J (1999) Restoration for soil protection after disturbances. In: Trabaud L (ed) Life and environment in Mediterranean biological systems, Computational Mechanics Publ, Southampton (in press)

Van der Drift JWM, Van Diepen CA (1992) The DBMETEO data base on the countries of the European Communities. Technical Document 4. SC-DLO Winand Staring Centre, Wageningen

Van der Voet P, Van Diepen CA, Voshaar JO (1994) Spatial interpolation of daily meteorological data. A knowledge-based procedure for the region of the European Communities. Report 53.3. DLO Winand Staring Centre, Wageningen

Van Wagner CE (1985) Drought, timelag and Fire Danger Rating. 8th National Conference on Fire and Forest Meteorology. Society of Amercian Foresters, Detroit, pp 178-185

Van Wagner CE (1987) Development and structure of the Canadian Forest Fire Weather Index System. Technical Report 35. Canadian Forestry Service, Ottawa

Van Wyngaarden R, Dixon R (1989) Application of GIS to model forest fire rate of spread. Challege for the 1990's GIS, Ottawa, pp 967-977

Vasconcelos MJ, Guertin DP (1992) FIREMAP. Simulation of fire growth with a Geographic Information System. International Journal of Wildland Fire 2: 87-96

Vasconcelos MJP, Paúl JCU, Silva S, Pereira JMC, Caetano MS, Catry FX, Oliveira TM (1998) Regional fuel mapping using a knowledge based system approach, In: Viegas DX, (ed), III International Conference on Forest Fire Research - 14th Conference on Fire and Forest Meteorology. ADAI, Luso-Coimbra, pp 2111-2123

Vázquez A, Moreno JM (1995) Patterns of fire occurrence across a climatic gràdient and its relationship to meteorological variables in Spain. In: Moreno JM, Oechel WC (eds) Global Change and Mediterrenean-Type Ecosystems, Springer-Verlag, New York, pp 408-434

Vega-García C, Woodard PM, Lee BS (1993) Mapping Risk of Wildfires from Human Sources of Ignition with a GIS. 13th Annual ESRI User Conference, pp 419-426

Vélez R (1993) High intensity forest fires in the Mediterranean Basin: Natural and socioeconomic causes. Disaster Management 5: 16-20

Vélez R (1996) Bilan des feux de forêt en Espagne en 1995 / En Espagne, pas de nouvelles, bonnes nouvelles. Forêt Méditerranéenne 17: 17-18 and 323

Vélez R (1997a) Recent history of fires in Mediterranean area. In: Balabanis P, Eftichidis G, Fantechi R (eds) Forest fire risk and management, Proceedings of the European School of Climatology and Natural Hazards course. European Commission, Brussels, pp 15-26

Vélez R (1997b) Les feux de forêt en Espagne en 1997. Forêt Méditerranéenne 18: 348

Vidal A, Pinglo F, Durand H, Devaux-Ros C, Maillet A (1994) Evaluation of a temporal fire risk index in Mediterranean forest from NOAA thermal IR. Remote Sensing of Environment 49: 296-303

Vidal A, Devaux-Ros C (1995) Evaluating forest fire hazard with a Landsat TM derived water stress index. Agricultural and Forest Meteorology 77: 207-224

Vidal A, Devaux-Ros C, Moran SM (1996) Atmospheric Correction of Landsat TM thermal band using Surface Energy Balance. Remote Sensing Reviews 15: 23-33

Vidal A, Desbois N, Pereira JMC, García J, Chuvieco E (1997) NOAA-AVHRR satellite data processing: state of the art and choices in Megafires, In: Chuvieco E (ed) A review of remote sensing methods for the study of large wildland fires. Universidad de Alcalá, Alcalá de Henares, pp 7-28

Viedma O, Melia J, Segarra D, Carcía-Haro J (1997) Modeling rates of ecosystem recovery after fires by using Landsat TM data. Remote Sensing of Environment 61: 383-398

Viegas DX (1998) Weather fuel status and fire occurrence: predicting large fires. In: Moreno JM (ed) Large Fires. Backhuys Publishers, Leiden, pp 31-48

Viegas DX, Viegas TP, Ferreira AD (1990) Characteristics of some forest fuels and their relation to the occurrence of fires. 1st International Conference of Forest Fire Research, Coimbra, pp B.03: 1-13

Viegas DX, Viegas TP, Ferreira AD (1991) Moisture content of fine forest fuels and fire occurrence in central Portugal. The International Journal of Wildland Fire 2: 69-85

Viegas DX, Bovio G, Camia A, Ferreira A, Sol B (1996) Testing Meteorological Fire Danger Methods in Southern Europe. 13th Conference on Fire and Forest Meteorology, Lorne, pp 571-589

Viegas DX, Piñol J, Viegas MT, Ogaya R (1998) Moisture content of living forest fuels and their relationship with meteorological indices in the Iberian Peninsula, In: Viegas DX (ed) III International Conference on Forest Fire Research - 14th Conference on Fire and Forest Meteorology. ADAI, Luso-Coimbra, pp 1029-1046

Viney NR (1991) A Review of Fine Fuel Moisture Modelling. International Journal of Wildland Fire 1: 215-234

Vitousek PM, Howarth RW (1991) Nitrogen limitation on land and in the sea: how can it occur. Biogeochemistry 13: 87-115

Vliegher BMd (1992) Risk assessment for environmental degradation caused by fires using remote sensing and GIS in a Mediterranean Region (South-Euboia, Central Greece). IGARSS'92. Int Geoscience and Remote Sensing Symposium, Houston, pp 44-47

Vliegher BMd, Dapper Md, Bazigos PS (1993) Assessment of fire risk in forests and grazing lands using multi source data. An example for SW-Messinia, Greece. In: van Genderen JL, van Zuidam RA, Pohl C (eds) International Symposium Operationalization of Remote Sensing. ITC, Enschede, pp 75-86

Wainwright J (1994) Anthropogenic factors in the degradation of semi-arid lands: a prehistoric case study in Southern France. In: Millington AC, Pye K (eds) Environmental changes in drylands: biogeographical and geomorphological perspectives. J Wiley & Sons, London, pp 427-441

Walker J, Raison RJ, Khanna PK (1986) Fire. In: Russell JS, Isbell, JS (eds) Australian soils. Univ Queensland Press, Australia, pp 185-216

Walsh SJ (1987) Comparison of NOAA-AVHRR data to meteorological drought indices. Photogrammetric Engineering and Remote Sensing 53: 1069-1074

Welch R, Ehlers W (1987) Merging multiresolution SPOT HRV and Landsat TM data. Photogrammetric Engineering and Remote Sensing 53: 301-303

Werle D, Drieman JA, Ahern FJ, O'Neil R (1991) Synthetic Aperture Radar (SAR): A promising tool for assessing recently burned forest in Canada? Proceedings of the 14th Canadian Symposium on Remote Sensing. Canadian Remote Sensing Society, Calgary, pp 470-475

Werth LF, McKinley RA, Chine EP (1985) The use of wildland fire fuel maps produced with NOAA-AVHRR scanner data. Pecora X Symposium, Fort Collins, pp 326-331

Westman WE, Price CV (1988) Spectral changes in conifers subjected to air pollution and water stress: experimental studies. IEEE Transactions on Geoscience and Remote Sensing 26: 11-20

Whelan RJ (1995) The Ecology of Fire. Cambridge University Press, Cambridge

Wiegand CL, Richardson AJ, Escobar DE, Gerbeman AH (1991) Vegetation indices in crop assessments. Remote Sensing of Environment 35: 105-119

Williams DL (1991) A comparison of spectral reflectance properties at the needle, branch and canopy level for selected conifer species. Remote Sensing of Environment 35: 79-93

Willis MJ (1985) Applications of Landsat imagery to fire fuels mapping projects covering large geographic areas. Pecora 10 Symposium. Remote Sensing in Forest and Range Resource Management, Fort Collins, pp 394-395

Woods JA, Gossette F (1992) A Geographic Information System for brush fire hazard management. ACSM/ASPRS Symposium, Washington, pp 56-65

Wooster MJ and Rothery DA (1997) Time-series analysis of effusive volcanic activity using the ERS Along Track Scanning Radiometer: the 1995 eruption of Fernandina Volcano, Galápagos Islands. Remote Sensing of Environment 62:109-117

Wukelic GZ, Gibbons DE, Martucci LM, Foote HP (1989) Radiometric calibration of Landsat Thematic Mapper thermal band. Remote Sensing of Environment 28: 339-347

Yool SR, Eckhardt DW, Estes JE, Cosentino MJ (1985) Describing the brushfire hazard in southern California. Annals of the Association of American Geographers 75: 417-430

Zhu Z, Evans DL (1994) U.S. forest types and predicted percent forest cover from AVHRR data. Photogrammetric Engineering and Remote Sensing 60: 525-531

Index